03-HHE-3

WORKSHOP ON MICROBIAL SOURCE TRACKING IN WATER

by:

Paul A. Rochelle
Ricardo De Leon
Metropolitan Water District of Southern California

2006

WERF

The Water Environment Research Foundation, a not-for-profit organization, funds and manages water quality research for its subscribers through a diverse public-private partnership between municipal utilities, corporations, academia, industry, and the federal government. WERF subscribers include municipal and regional water and wastewater utilities, industrial corporations, environmental engineering firms, and others that share a commitment to cost-effective water quality solutions. WERF is dedicated to advancing science and technology addressing water quality issues as they impact water resources, the atmosphere, the lands, and quality of life.

For more information, contact:
Water Environment Research Foundation
635 Slaters Lane, Suite 300
Alexandria, VA 22314-1177
Tel: (703) 684-2470
Fax: (703) 299-0742
www.werf.org
werf@werf.org

This report was co-published by the following organization. For non-subscriber sales information, contact:

IWA Publishing
Alliance House, 12 Caxton Street
London SW1H 0QS, United Kingdom
Tel: +44 (0) 20 7654 5500
Fax: +44 (0) 20 7654 5555
www.iwapublishing.com
publications@iwap.co.uk

Library of Congress Catalog Card Number: 2005938196
IWAP ISBN: 978-1-84339-745-8

Metropolitan Water District of Southern California

ACKNOWLEDGMENTS

The authors wish to thank all individuals who participated in the workshop, particularly those who agreed to make presentations and lead breakout groups (indicated with an asterisk in the list of participants). Peter Hartel's thorough editorial review of the material in Chapter 1.0 is greatly appreciated. The Principal Investigators are grateful to Patrick Hacker at MWDSC for stepping in as a co-facilitator at the workshop with only a few days notice. In addition, Lynn Kelemen, Susan Hogan, and Dorothy Neault provided invaluable assistance with administrative issues prior to and after the workshop. They are also grateful to Mark Tenneson for making most of the travel arrangements for the project team and participants. Finally, the input of the project subcommittee and project manager is appreciated.

Report Preparation

Principal Investigators:
Paul A. Rochelle
Ricardo De Leon
Metropolitan Water District of Southern California

Project Team:
Patrick Hacker
Metropolitan Water District of Southern California

Kate Martin
Independent Consultant

Project Subcommittee

Alfred Dufour
U.S. Environmental Protection Agency

Michael Jenkins
U.S. Department of Agriculture

Sharon C. Long
University of Massachusetts Amherst

Charles McGee
Orange County Sanitation District

Margaret Nellor
Sanitation Districts of Los Angeles County

Jon Standridge
Wisconsin State Laboratory of Hygiene

Scott Woodard
CTE Engineers, Inc.

Water Environment Research Foundation Staff

Director of Research: Daniel M. Woltering, Ph.D.
Project Manager: Jami Montgomery

Workshop Participants

Nicholas Ashbolt*
The University of New South Wales

Mark Borchardt
Marshfield Clinic – Marshfield Center

George Di Giovanni
The Texas A&M University System

Tom Edge*
Environment Canada

Rudolph Ellender
University of Southern Mississippi

Katharine Field
Oregon State University

David M. Gordon*
The Australian National University

William Hamilton
Vanderbilt University

Peter G. Hartel*
University of Georgia

Valarie J. Harwood
University of South Florida

Randall Hicks*
University of Minnesota-Duluth

Kenneth Hyer*
U.S. Geological Survey

James Johnson
Minneapolis VA Medical Center

James Kern
Virginia Tech Corporate Research Center

Christopher L. Kitts
California Polytechnic San Luis Obispo

Menu Leddy
Orange County Water District

Jennifer MacDonald
University of Georgia

Sandra McLellan*
Great Lakes WATER Institute

Richard J. Meinersmann
U.S. Department of Agriculture

Douglas Moore
Orange County Public Health Laboratory

Sam Myoda
Delaware Natural Resources and Environment Control

Cindy H. Nakatsu
Purdue University

Robin Oshiro*
U.S. Environmental Protection Agency

Jeffrey L. Ram
Wayne State University

Mansour Samadpour*
Institute for Environmental Health

Jorge Santo Domingo*
U.S. Environmental Protection Agency

Marion Savill
Institute of Environmental Science and Research, New Zealand

Geoffrey Scott*
National Oceanic and Atmospheric Administration

Marilyn Soule
Washington State University

Jill Stewart
National Oceanic and Atmospheric Administration

Huw Taylor*
Brighton University, UK

Jan Vinje
University of North Carolina

Stephen Weisberg*
Southern California Coastal Water Research Project Authority

Bruce Wiggins
James Madison University

Jayson Wilbur
Worcester Polytechnic Institute

ABSTRACT AND BENEFITS

Abstract:

The drinking water and wastewater industries are interested in developing a better understanding of sources of fecal contamination. Microbial source tracking (MST) offers the potential to apportion the contribution of various animal groups to this contamination. However, while there are many methods available, there is no clear indication as to the most appropriate approach. There has been little systematic comparison of methods and only a few blind trials, and issues of reproducibility, accuracy, sensitivity, specificity, resolution, and robustness need to be addressed. There is confusion among potential end-users and some disagreement among scientists. Therefore, due to the uncertainty surrounding the most appropriate tools and applications of MST, a workshop of 45 experts representing water and wastewater utilities, academia, state and federal government agencies, medical institutions, and private laboratories was convened in San Antonio, Texas over a three-day period (February 16-18, 2005). The overall objective of the workshop was to identify the knowledge gaps and research needs for application of MST technologies by the wastewater and drinking water industries. The general conclusion was that there is a role for MST so long as it is part of an overall strategy that includes: multiple congruent methods; traditional land use, sanitary, and wildlife surveys; the geographic scale is not too large; and so long as defined criteria of standardization, quality assurance and quality control (QA/QC), and measures of data quality are met. However, without further improvements in method accuracy and robustness, it is unlikely that any of the proposed methods can reliably identify fecal contamination at the level of individual host animals.

While many issues were discussed during the workshop, some topics became recurring themes over the three-day period and are therefore seen as some of the most critical areas for consideration. There needs to be more inter-laboratory comparison of methods, particularly method combinations, generation of and subsequent adherence to standardized procedures, greater inclusion of QA/QC aspects into MST studies, increased publication of MST success stories, and more thorough documentation of all study aspects in publications. It was recognized that some methods have achieved a sufficient level of maturity such that they can be considered as candidates for standardization, whereas others are still experimental or research grade tools. Cost of analysis was a recurring theme throughout the workshop with some calls to develop reliable but inexpensive methods rather than increasingly sophisticated techniques requiring specialized equipment and the development of huge libraries of strains. There was considerable debate regarding adopting a pragmatic approach, where studies are limited by funding and timeline, and human versus non-human discrimination is adequate, as opposed to more idealistic methods involving libraries containing many thousands of isolates, techniques with increasing complexity, increased certainty in the results, and higher levels of discriminatory resolution. Since MST tools may be used more routinely in the future to help utilities comply with TMDL regulations and other application drivers, a pragmatic approach to their application will need to be adopted. If their regulation-driven application cannot wait for the science to mature sufficiently, then it will be necessary to incorporate a higher level of QA/QC into MST studies.

There has been considerable debate over which method is the "best", but a consensus has been reached that there is unlikely to be a single "silver bullet". Rather, MST end-users should utilize a toolbox approach, the individual components of which will vary depending on the type of watershed, potential sources of contamination, funding and time available, and the level of discriminatory resolution required. Future comparison studies should evaluate different suites of methods to determine those that best meet accuracy, sensitivity, and reproducibility criteria. In addition to conducting the best possible science to further develop and evaluate the science of MST methods, there is a need for public outreach so that MST customers, regulatory agencies, utility managers, and other end-users have realistic expectations about what can, and cannot, be achieved by MST. Whichever approach is adopted, utilities need to have a clear understanding of the questions being asked, the expected level of resolution, an acceptable level of ambiguity, what kind of results to expect, how to interpret results, and the costs involved. In addition, MST should not be the only tool used to determine the source(s) of fecal contamination. Fecal source tracking programs also need to include watershed-specific studies of agricultural land use management practices, surveys of sewage and septic systems, stream/river flow patterns, seasonal run-off patterns, sewage treatment practices, and wildlife surveys.

Benefits and Recommendations:

The outcome of the workshop benefits the wastewater and drinking water industries, and other potential end-users, by providing a framework for the future development and improvement of MST technologies. Some of the recommendations that were developed during the workshop relate to broad policy issues such as increased funding and improved communication. Others deal with specific research topics, some of which will require long-term programmatic funding efforts while others may be accomplished in a single project. Specific recommendations include:

- Since most MST technologies are not yet ready to be used in support of management decisions, the wastewater and drinking water industries should exercise caution when considering applications of MST.
- Creation of a long-term, multi-agency program for funding MST research is necessary to enable method comparisons and field studies of sufficient depth and scope.
- Increased public outreach is required to ensure that potential end-users have realistic expectations about what can be achieved with MST technologies.
- Establishment of a comprehensive FST database providing information on the status, relative costs, advantages, and limitations of methods, availability of contract laboratories, location of MST researchers, and links to peer reviewed literature.
- A series of inter-laboratory comparisons of various method combinations (toolbox approach) should be performed.
- Field studies should be performed using a combination of at least two of the best available methods before and after remediation activities.
- MST methods need to be standardized and must incorporate QA/QC, and appropriate criteria must be developed to allow this to occur.
- More research is necessary on geographic and spatial stability of target organisms and markers, inter-laboratory reproducibility, quantification methods, and investigation of chemical methods.

- The fate and persistence of fecal indicators in the environment should be investigated.
- The feasibility of merging existing databases for library-based methods and making them publicly available should be assessed.
- Unresolved questions concerning species heterogeneity and growth and persistence in the environment demonstrate the necessity of developing alternatives to *Escherichia coli* and enterococci as MST target organisms.

It is suggested that the most useful role for WERF in the ongoing development of MST is to sponsor laboratory and field-based validations and inter-laboratory comparisons of various method combinations (the toolbox approach) while promoting the development of standardized methods and QA/QC measures. Research that leads to the development of a toolbox approach that can accurately, reliably, and consistently discriminate between human and non-human contamination should be funded. This toolbox must be supported by extensive evaluation and pilot-scale field trials, inter-laboratory comparison studies demonstrating high reproducibility, and independent validation. These goals may best be achieved by partnering with other organizations in developing a program for pragmatic and defensible applications of MST.

Keywords: Agriculture, antibiotic resistance, bacteriophage, best management practices, enterococci, *Escherichia coli*, fecal contamination, fecal sterols, human waste, livestock, microbial fingerprinting, microbial indicators, PFGE, polymerase chain reaction, ribotyping, standardization, statistical analysis, wildlife.

TABLE OF CONTENTS

LIST OF TABLES

LIST OF FIGURES

LIST OF ACRONYMS

AFLP	Amplified fragment length polymorphism
ARA	Antibiotic resistance analysis
ARCC	Average rate of correct classification
ARP	Antibiotic resistance profiling (syn. ARA)
ASTM	American Society for Testing and Materials
BMP	Best management practices
BOX-PCR	Repetitive polymerase chain reaction using BOX primers
BST	Bacterial source tracking
CCR	Correct classification rate
CDC	Centers for Disease Control and Prevention
CSU	Carbon source utilization (syn. CUP)
CUP	Carbohydrate utilization profiling
DA	Discriminant analysis
DFA	Discriminant function analysis (syn. DA)
DGGE	Denaturing gradient gel electrophoresis
DNA	Deoxyribonucleic acid
EcoRI	Restriction endonuclease derived from *Escherichia coli*
EfOM	Effluent-derived organic matter
ERIC-PCR	Enterobacterial repetitive intergenic consensus sequences polymerase chain reaction
ETV	Environmental Technology Verification
FDA	Food and Drug Administration
FST	Fecal source tracking
GC/MS	Gas chromatography/mass spectrometry
HindIII	Restriction endonuclease derived from *Haemophilus influenzae*
ISR-PCR	Intergenic spacer region polymerase chain reaction
MALDI-TOF	Matrix-assisted laser desorption ionization-time of flight mass spectrometry
MAR	Multiple antibiotic resistance (syn. ARA)
MDP	Minimum detectable percentage
MLEE	Multilocus enzyme electrophoresis
MLST	Multilocus sequence typing
MRA	Multiple resistance analysis (syn. ARA)
mtDNA	Mitochondrial DNA
MST	Microbial source tracking
PCR	Polymerase chain reaction
PFGE	Pulse field gel electrophoresis
PSC	Project Scientific Committee
PvuII	Restriction endonuclease derived from *Proteus vulgaris*
QA/QC	Quality assurance and quality control
Q-PCR	Quantitative PCR
rep-PCR	Repetitive polymerase chain reaction

REP-PCR	Repetitive extragenic palindromic sequence polymerase chain reaction
RFLP	Restriction fragment length polymorphism
RFP	Request for proposals
rRNA	Ribosomal ribonucleic acid
SOP	Standard operating procedure
TMDL	Total maximum daily load
TRFLP	Terminal restriction fragment length polymorphism
USDA	United States Department of Agriculture
U.S. EPA	United States Environmental Protection Agency
USGS	United States Geological Survey
WERF	Water Environment Research Foundation

EXECUTIVE SUMMARY

ES.1 Background

Microbial source tracking (MST) technologies offer the potential to identify sources of fecal contamination in water. Such information is critically important to the wastewater, stormwater, and drinking water industries. One of the primary drivers for moving MST technologies out of a research setting and into "real-world" applications in the U.S., even though most of the technologies are not sufficiently mature, is requirements of the Total Maximum Daily Load (TMDL) regulations to account for all sources of pollutants. A TMDL is a qualitative assessment of water quality problems, contributing sources of contamination, and load reductions or control actions that are necessary to restore and protect individual bodies of water. Sources of contamination may be point and non-point human sources as well as naturally occurring non-human sources.

There are many proposed MST methods encompassing chemical and biological targets, phenotypic and genotypic techniques, culture-dependent and independent methods, and library-dependent and independent methods. Since some methods target chemicals rather than microbes, a more appropriate terminology is fecal source tracking (FST). In general, FST methods are not well established and their adoption by non-specialist laboratories would require substantial training and resources. Most of the methods are research-grade tools that are not yet ready for routine use in utility decision making processes. There have been attempts to standardize and validate some protocols, but these protocols are not widely available. There has been little systematic comparison of methods and issues of reproducibility, accuracy, sensitivity, resolution of source animal discrimination, method robustness, and methods of data analysis need to be addressed. In addition, issues of quality control and quality assurance (QA/QC) have not been addressed in sufficient detail.

Therefore, a workshop on MST was convened in San Antonio, Texas during February 2005. The primary objective for the workshop was to identify the knowledge gaps and research needs for applying MST technologies by the wastewater and drinking water industries. Participants in the workshop included 47 individuals (including the seven member project scientific committee and WERF project manager) from water and wastewater utilities, universities, state and federal government environmental and agricultural agencies, medical institutions, and private laboratories. The workshop was also attended by 75 observers representing local (Texas) water agencies, state government agencies, commercial laboratories, and academia. Technical presentation topics included a report from a 2002 MST workshop, the assumptions and limitations of MST, the confounding role of sediment in MST, application of culture-independent methods, population genetics and ecology of *Escherichia coli*, comparison of methods, applications of library-independent chemical FST methods, case studies in marine and freshwater systems, and an overall European perspective of MST. Breakout groups discussed issues centered on the selection of the most appropriate organisms and targets for MST, study design criteria and standardization, information handling and data analysis. Discussions throughout the workshop included the appropriateness of *E. coli* and enterococci as targets for MST, whether there is a future for library-based methods, the ideal library size, the measures of data quality that should be reported in the literature, how to assess the performance and accuracy of methods, and the importance of quantification for MST methods.

ES.2 Conclusions

The general conclusion from the workshop was that there is almost certainly a future role for MST in the wastewater and drinking water industries, so long as it is part of an overall strategy that includes multiple congruent methods, traditional inquiries (land use, sanitary, and wildlife surveys), the geographic scale is not too large, and so long as defined criteria of standardization, QA/QC, and measures of data quality are met. However, without further improvements in method accuracy and robustness, it is unlikely that any of the proposed methods can reliably identify fecal contamination at the level of individual host animals. There are legal, political, and financial consequences of MST-derived conclusions and remediation programs. Therefore, variability, reproducibility, and reliability are all critical aspects of future MST development because researchers may find themselves in court defending MST methodology when used to inform TMDL or other decision making processes. Consequently, the MST methods selected must be scientifically sound and legally defensible.

Participants agreed that more comparison and evaluation studies need to be conducted. There needs to be more inter-laboratory comparison of methods, generation and adherence to standardized procedures, greater inclusion of QA/QC into MST studies, increased publication of MST success stories, and more thorough documentation of all study aspects in publications. It was recognized that some methods have achieved a sufficient level of maturity such that they can be considered as candidates for standardization, whereas others are still experimental or research grade tools. Cost of analysis was a recurring theme throughout the workshop with some calls to develop reliable but inexpensive methods rather than increasingly sophisticated techniques requiring specialist equipment and the development of huge libraries of strains. There was considerable discussion of the need to adopt a pragmatic approach, where studies are limited by funding and timeline and human versus non-human discrimination is adequate, as opposed to more idealistic methods involving libraries containing many thousands of isolates, techniques with increasing complexity, increased certainty in the results, and higher levels of source discrimination. Since MST tools are likely to be used more routinely in the future to help utilities comply with TMDL regulations and other application drivers, a pragmatic approach to their application will need to be adopted.

In response to a suggestion that recent method comparison studies have eroded the confidence in current MST methods, participants responded that they have in fact increased the credibility of the field since researchers are recognizing and publishing the limitations of the methods. It was recognized that conducting science in the public eye does not always cast the results in a favorable light. Therefore, in addition to conducting the best possible science to further develop and evaluate the science of MST methods, there is a need for public outreach so that MST customers, regulatory agencies, public utilities, and other end-users have realistic expectations about what can, and cannot, be achieved by MST. Continued communication within the research community and improved communication between researchers and end-users is vital for the future success of MST technologies.

Workshop participants reached the following specific conclusions regarding critical issues that must be addressed if MST is to continue moving forward:

- There is insufficient information to discount any of the current methods from future consideration. However, the "silver bullet" ideal method probably does not exist; no single technique will be suitable in all situations. Therefore, a toolbox approach involving multiple complementary methods is better suited for field applications. The components of this toolbox will vary depending on the:
    - Geographical location, limnology, hydrology, and physical characteristics of the watershed or body of water under investigation.
    - Diversity and number of potential source(s) of contamination.
    - Required or acceptable level of source discrimination.
    - Required level of certainty in the results.
    - Level of funding available.
    - Urgency to generate results.
- Recent method comparison studies have given researchers a "reality check" on the performance of MST methods. It is now necessary to communicate this to customers in terms of realistic expectations and limitations. In addition to conducting the best possible science to further develop and evaluate MST technologies, there is a need for public outreach so that potential customers know what can, and cannot, be achieved by MST.
- Although there are reports that describe source animal discrimination at the level of individual host species, recent comparison studies cast doubt on the ability of any of the techniques to achieve this resolution on a consistent basis. However, it can be concluded that some MST methods can reliably discriminate between human and non-human sources of contamination.
- At a utility management level, discrimination between human, livestock, and wildlife sources of contamination will usually be sufficient. However, there will be circumstances where three-way discrimination is insufficient and the end-user needs to identify individual types of contributing animal.
- More inter-laboratory, multi-investigator comparison studies and evaluation of methods are needed. While evaluation of individual methods is still necessary to assess their source prediction accuracy, reproducibility, sensitivity, and specificity, the primary focus of future investigations should be comparisons between various combinations of methods.
- There is an immediate need for MST methods to be standardized if the techniques and their results are to withstand scientific scrutiny and legal challenges. Once standardized, published or otherwise publicly available SOPs must be adopted and followed by anyone performing a particular technique, if the results are to be given serious consideration by the wider MST and end-user communities.
- Incorporation of QA/QC measures into all aspects of MST research and application is critical.
- Performance evaluation criteria need to be developed so that all researchers use the same approach when developing, evaluating, and comparing methods. Also, it may be necessary to establish a mechanism for independent oversight of method evaluations or for independent testing of methods.

- Further development of new methods is necessary. Such methods may include multi-target microarrays, eukaryotic mitochondrial DNA, new host-specific organisms or molecular markers in specific organisms. Metagenomics should also be investigated as a process for target selection.
- Improved documentation of all aspects of MST research and field applications is necessary, including more case study publications, particularly increased dissemination of success stories, more detailed method descriptions, documentation of QA/QC measures, and the steps taken to assess data quality.
- A mechanism is needed to promote the transition of sufficiently standardized and validated methods from a research setting into real-world applications.
- The design of MST field application studies needs to consider appropriate sampling regimes, including sample site locations, number and frequency of samples, number of isolates, types and number of potential source animals, number of isolates per animal, and number of replicates per isolate.
- Selecting appropriate data analysis tools is seen as a very important issue since there is growing concern about getting different MST answers depending on the method of data analysis that was used. Research is needed to determine the most apposite similarity threshold for band-matching methods such as rep-PCR, ribotyping, and PFGE.
- A greater understanding is needed of the relationship between watershed land use and the sources of bacteria and/or indicators in streams. Research also needs to be conducted on the potential confounding role of indicators in sediments.
- Mechanisms should be established to increase sharing of data, possibly through the creation of publicly available databases. Also, a comprehensive FST database would benefit the entire end-user community. Information to include in such a database includes the status, relative costs, advantages, and limitations of various methods, availability of contract laboratories, location of MST researchers, and links to peer reviewed literature.
- MST studies should focus on locations where remediation efforts are planned so that results prior to, and following remediation can be compared. MST should only be considered successful if the resulting remediation measures are successful (i.e., contamination is reduced).
- Funding of MST-related research and of field applications is a critical issue. There must be sufficient funding on the research side to allow implementation of standardization, QA/QC, rigorous testing, method comparisons, and development of new methods. Also, field studies must be adequately funded to allow sufficient samples to be analyzed with the most appropriate methods. Therefore, outreach is required to funding organizations, government agencies, the wastewater and drinking water industries, and other end-users, to promote the availability of adequate resources.

ES.3 Recommendations

Some of the recommendations that were developed during the workshop relate to broad policy issues such as increased funding and improved communication. Others deal with specific areas where the science of MST needs to progress, some of which will require long-term programmatic funding efforts while others may be accomplished in a single project.

ES.3.1 Policy Recommendations

- Some current methods can potentially be used for distinguishing human from non-human sources of contamination in small-scale field studies. However, the wastewater and drinking water industries should exercise caution when considering MST applications. The technologies are not yet ready to be used reliably and consistently for supporting watershed management or TMDL-related decisions. However, despite this lack of maturity with the methods, they are likely to be used with increasing frequency over the next few years. Consequently, a useful role for interested organizations such as WERF, in the ongoing development of MST, would be to sponsor laboratory and field-based validations and inter-laboratory comparisons of various method combinations (the toolbox approach) while promoting the development of standardized methods and QA/QC measures.
- Creation of a long-term, multi-agency program is necessary for funding MST research through a mechanism that facilitates cooperative funding between agencies with a stake in the development and successful application of MST. Such agencies might include WERF, U.S. Environmental Protection Agency, Awwa Research Foundation (AwwaRF), Environment Canada, National Oceanic and Atmospheric Administration, U.S. Geological Survey, and U.S. Dept. of Agriculture.
- Organization of an MST customer workshop aimed at instilling realistic expectations so that end-users are aware of what is, and what is not, possible with current techniques. Once initiated, this communication should continue so that end-users are provided with regular updates on the progress of the science as well as successes and problems arising with field applications.
- Coordination between the organizers of future MST workshops and meetings should be encouraged so that each new gathering of experts builds upon the experiences and lessons learned from earlier efforts.

ES.3.2 Research Recommendations

- In a move towards a toolbox approach, there is a need to conduct comparison studies designed to evaluate different suites of methods to determine those that best meet accuracy, sensitivity, and reproducibility criteria. Inter-laboratory comparisons of various method combinations incorporating standardization, QA/QC, and *a priori* criteria for assessing method performance, are necessary. These studies should also assess the different statistical methods that could be used to analyze the combined dataset.
- It is necessary to determine the most appropriate approach for standardizing methods and incorporating elements of QA/QC into MST studies. Criteria for acceptable performance of a method must be developed along with a strategy for testing these criteria. Submission of methods for independent evaluation using the predetermined criteria and testing strategy is recommended.
- Field studies using a combination of at least two of the best available methods before and after remediation activities will help to determine whether remediation efforts are successful and also if FST methods can detect (and measure) the affect of remediation.
- More specific research is necessary for many of the methods, including:

- Geographical specificity and inter-laboratory reproducibility of host-specific gene markers need to be assessed along with an evaluation of the level of source discrimination and quantification that are feasible.
    - More rigorous testing of library-based methods is required, including developing tools that can be used to determine the appropriate size of the library.
    - Spatial and temporal stability of all markers needs to be investigated more thoroughly.
    - Further investigation of chemical methods is necessary to address questions concerning persistence, biodegradability, and correlation to microbial indicators.
    - Standardized controls need to be developed for all PCR-based methods.
- The natural history of fecal indicators in the environment should be investigated, including assessing the short- and long-term fate, ecology, and population genetics of all target organisms and markers.
- A study is needed to assess the feasibility of merging existing databases for library-based methods and making them publicly available.
- Research is needed to discover alternatives to *E. coli* and enterococci as MST target organisms since intra-species heterogeneity, growth and persistence of these bacteria in the environment, and lack of animal host specificity may limit their utility.

For potential end-users within the drinking water and wastewater industries attempting to identify sources of fecal contamination using an FST approach, the options that are currently available include: developing in-house capabilities; submitting isolates to a commercial service to perform the FST analyses; contracting out the entire investigation; or collaborating with individual method developers in academic institutions. Whichever approach is adopted, utilities need to have a clear understanding of the questions being asked, the expected level of resolution, an acceptable level of ambiguity, what kind of results to expect, how to interpret results, and the costs involved. Issues of quality assurance, method reproducibility, and statistical robustness should also form part of the discussion. In addition, FST should not be the only tool used to determine the source(s) of fecal contamination. Fecal source tracking programs also need to include watershed-specific studies of agricultural land use management practices, surveys of sewage and septic systems, stream/river flow patterns, seasonal run-off patterns, sewage treatment practices, and wildlife surveys.

ES.3.3 Recommendations for WERF

The following prioritized recommendations highlight a potential role for WERF in advancing the application of FST technologies by ensuring the highest quality practical science while maintaining realistic expectations:

1. Sponsor inter-laboratory comparisons and field-based trials of various method combinations (the toolbox approach) while promoting the development of standardized methods and QA/QC measures.

2. Fund research that leads to the development of a toolbox approach that can accurately, reliably, and consistently discriminate between human and non-human contamination. This toolbox must be supported by extensive evaluation and pilot-scale field trials, inter-laboratory comparison studies demonstrating high reproducibility, and independent validation.
3. Partner with other organizations in developing a program for pragmatic and defensible applications of FST.
4. Sponsor public outreach events aimed at instilling realistic expectations for potential MST customers.

CHAPTER 1.0

INTRODUCTION: STATUS OF MICROBIAL SOURCE TRACKING METHODS

1.1 Background

The summary on the status of microbial source tracking (MST) methods in this chapter was written prior to the San Antonio workshop as preparation material for the project scientific committee (PSC) and workshop participants. It was reviewed by the PSC and some participants. For the preparation of this report, it has been revised in line with comments received during and since the workshop. It was not intended as an exhaustive review of all source tracking methods and studies, but rather as a starting point for the workshop.

Within the drinking water industry, protecting source water from contamination by pathogens is seen as increasingly important for public health. Similarly, the wastewater industry is concerned about protecting beneficial uses of receiving waters that are used for drinking water and/or recreation, and the contributions of discharges and overflows to total maximum daily loads (TMDL). Receiving waters for one utility may be source water for a second, and both drinking and wastewater industries have common concerns about contamination. Potential sources of fecal contamination include runoff from concentrated animal feedlot operations, manure- or biosolids-amended agricultural lands, urban run-off that may reach receiving waters as discharges from storm sewers or as combined sewer overflows, malfunctioning septic systems, domestic animals, and wildlife. Urban wet weather pollution is a major source of impairment of water quality in many receiving waters. Sources of bacteria in storm water include domestic pets, urban wildlife, human waste via storm and sanitary sewer cross-connections, rodents dwelling in sewers, and growth of bacteria in nutrient-rich water standing in storm sewers between events. According to the U.S. Environmental Protection Agency (U.S. EPA), over 40% of assessed waters do not meet the quality standards set by various legislative bodies. Consequently, preventing contamination is a primary goal of both industries. However, preventative measures are difficult to apply if the source of contamination cannot be identified. Although standard microbiological assays such as detection of fecal coliforms, *Escherichia coli*, and enterococci are used as indicators of microbiologically impaired water quality, they do not provide information on the source of contamination. The ability to differentiate between human and non-human sources of fecal contamination, or between domestic and wild animals, or between different species of animals, would significantly enhance programs aimed at protecting source and receiving waters from contamination.

A wide variety of methods has been proposed for identifying the source of fecal contamination in water (Table 1-1). Chemical methods such as the detection of detergents and optical brightening agents, triclosan, caffeine, prescription and non-prescription drugs, and sterols assume an exclusively anthropogenic source for these compounds. A variety of microbiological and molecular methods have also been proposed to determine the human and non-human hosts of fecal bacteria and thus to track the source of contamination in water. Molecular methods are potentially powerful tools that can discriminate between bacteria at the species and subspecies levels. Some of these methods use bacterial species that are specific to particular animal hosts, while others focus on sub-typing particular species on the assumption that host-adapted strains can be identified.

Table 1-1. Potential Fecal Source Tracking Methods.

A. Library Independent Methods	B. Library Dependent Methods
Host-specific organisms	**Genotypic methods**
• *Bacteroides fragilis* phages	• Amplified fragment length polymorphism (AFLP)
• Coliphages	• Denaturing gradient gel electrophoresis (DGGE)
• *Bifidobacterium* spp.	• Multilocus sequence typing (MLST)
• *Clostridium perfringens*	• Plasmid typing
• *Fusobacterium*	• Pulsed field gel electrophoresis (PFGE)
• *Peptococcus*	• Random amplification of polymorphic DNA (RAPD)
• *Rhodococcus coprophilus*	• Repetitive element polymerase chain reaction (REP-, BOX-, ERIC-PCR)
• *Ruminococcus*	• Restriction fragment length polymorphism (RFLP)
• *Streptococcus bovis*	• Ribotyping
• *Enterococcus faecalis*	• Terminal restriction fragment length polymorphism (tRFLP)
• Fecal bacteria ratios	• Sequencing organism-specific genes
• Human/animal-specific viruses	
Genotypic methods	**Phenotypic methods**
• *Bacteroides-Prevotella* spp.	• Antibiotic resistance analysis (ARA)
• Coliphage genotyping	• Carbohydrate utilization profiling (CUP)
• Enterotoxin biomarkers	• *Escherichia coli* serotyping
• *Enterococcus faecium* esp gene	
• *Cryptosporidium/Giardia* genotyping	
Chemical methods	
• Non-prescription drugs	
• Caffeine	
• Fecal sterols and stanols	
• Cotinine	
• Detergents	
• Fluorescent whitening agents	
• Triclosan	
• Fatty acids	
• Polar lipids/whole cells (by MALDI-TOF)	

Microbial source tracking (MST) has the potential to improve the management of wastewater discharges, agricultural run-off, and source water protection. The drivers for the application of MST technologies include:

- Regulations such as the Clean Water Act.
- Beach closures (coastal and freshwater) affecting recreation-related revenues.

- Source protection for drinking water.
- TMDL programs.
- Shellfish harvesting.
- Livestock health.
- Agriculture encompassing the effects of agriculture and the effects on agriculture.

However, many potential source tracking methods have been described and there is currently no consensus on the most appropriate technique(s) to use in actual source tracking investigations. Typically, microbial methods are divided into library dependent and library independent techniques (Table 1-1). Methods can also be divided based on the necessity for culturing specific microorganisms versus culture-independent techniques, or they can be divided into phenotypic or genotypic methods. Although there are supporters for each of the available technologies, many questions about these technologies remain unanswered. There are many examples of various source tracking tools in real world situations to determine the source of contamination in freshwater and marine environments. While some of these studies were able to determine the animal source of contamination with reasonable certainty, many studies were unable to determine the specific culprit(s). Despite the many reported successes of the application of MST methods, case studies and reviews frequently draw conclusions such as:

- "The accuracy of bacterial source tracking methods as source assessment tools has not been well established".
- "There is no easy, low-cost method for differentiating between human and non-human sources of contamination".
- "Quantifying the contribution from different sources is not possible".
- ".......the results indicated that ARA and ribotyping currently lack the accuracy and reproducibility required to determine the sources of bacterial pollution".
- "Further studies are needed to validate source tracking methods using quality assurance testing".

Consequently, a synthesis of the current state of knowledge is required for fecal source tracking (FST) technologies. In the last few years there have been numerous reviews on the state of the science of source tracking methods (Field, 2004; Rangdale, 2003; Scott et al., 2002; Simpson et al., 2002; Sinton et al., 1998), method comparisons (Carson et al., 2003; Field et al., 2003; Griffith et al., 2003; Moore et al., 2005; Myoda et al., 2003) as well as many unpublished reviews and method critiques. There have also been many workshops and conference symposia dedicated to MST in recent years (see Chapter 7.0). In addition, there are numerous websites dedicated to fecal source tracking (Chapter 7.0), the most informative of which can be accessed through the U.S. Geological Survey website- http://water.usgs.gov/owq/microbial.html. Bacterial source tracking is also been applied as one of the approaches in a number of multi-state projects aimed at improving environmental and public health, such as the U.S. EPA Gulf of Mexico Program (http://www.epa.gov/gmpo/) and Multi-State Research Project S-297 (http://s-297.okstate.edu). The advantages and limitations of each of the available methods need to be understood by utilities and other end-users, along with the most appropriate application of the technologies. Selecting an appropriate source tracking strategy will not only be based on objective measures of accuracy and precision of a particular method or suite of methods, but also on cost of analysis, required level of sophistication, availability of in-house or commercial expertise, time-frame, and the number and type of potential sources of contamination.

1.2 Chemical Methods

A variety of chemical compounds has been proposed as potential FST tools (Table 1-1). Caffeine, pharmaceuticals, fluorescent whitening agents, and various anti-microbial agents are all excreted or disposed of in large quantities by modern human societies. Coprostanol (5β-stanol) is a derivative of cholesterol formed by indigenous intestinal bacteria in the human gut and accounts for approximately 60% of the total sterols in human feces (Leeming et al., 1996). The concentration of coprostanol in human feces is approximately 10-fold higher than in the feces of other animals while the dominant sterol in herbivores is 24-ethylcoprostanol (Leeming et al., 1996). Low levels of coprostanol are also excreted by a variety of animals including cows, horses, possums, pigs, and cats. The sterol content of bird feces is extremely variable. Coprostanol may be positively correlated with fecal coliforms, *Clostridium perfringens* spores, and *E. coli* in freshwater and coastal water (Edwards et al., 1998; Nichols et al., 1993; Isobe et al., 2004). Analysis of a panel of seven fecal sterols indicated contamination of catchment waters by dog, bird, and human feces (Suprihatin et al., 2003). The ratio of coprostanol to predominant herbivore stanols was combined with the detection of *E. coli*, *Bifidobacterium adolescentis*, *Rhodococcus coprophilus*, and human-specific *Bacteroides* spp. to identify human-derived fecal pollution in river water (Gilpin et al., 2003). Also, Puglisi et al. (2003) detected coprostanol in all sewage-amended soils and soil contaminated with municipal waste, but not in untreated soils. Therefore, this compound may be a good marker of human fecal contamination. In contrast, various authors have reached opposite conclusions regarding the correlation between fecal sterols and microbial indicators of fecal contamination. These authors contend that coprostanol is present in other animals and its detection cannot be categorical evidence of human contamination. Hyodeoxycholic acid, a bile acid characteristic of pig feces, has been suggested as a pig-specific indicator (Elhmmali et al., 1997). A combination of bile acids, stanols, and sterols was used to assess fecal contamination in river water (Elhmmali et al., 2000). The concentrations of fecal bile acids showed strong agreement with coprostanol concentrations. This multi-molecule approach may be better to assess sources of fecal input.

The average concentrations of a range of steroid compounds in a California watershed ranged from 2 ng/L for epicoprostanol to 811 ng/L for cholesterol (Noblet et al., 2004). The lack of a significant correlation between coprostanol concentrations and concentrations of fecal indicator bacteria (enterococci, total coliforms, and *E. coli*) suggests that raw sewage and treated wastewater effluent were not contributing fecal contamination to the river. Significant correlations between bird signature steroids and fecal indicator bacteria implicated birds as a potential source of contamination.

Human effluent in the form of community wastewater and septic tanks could be distinguished from non-human effluents by the presence of fluorescent whitening agents (Gregor et al., 2002). However, detergents and whitening agents have not been directly linked to fecal pollution and may at best only point to industrial contamination or general human activity.

One assumption behind many of the chemical-based methods is that human activity is the only source of these compounds; therefore, detecting them in water indicates an anthropogenic source of contamination. However, many of the compounds occur widely in environmental waters, which limit their use for direct source tracking studies. A survey of 139 streams across 30 U.S. states reported detection frequencies of 86%, 71%, and 58% for coprostanol, caffeine, and triclosan, respectively, with maximum concentrations of 2-10 μg/L (Koplin et al., 2002). Also, chemical methods may lack the resolution of microbial-based methods and can generally only be used at a coarse resolution level to determine whether an anthropogenic contamination source exists. In addition, the issue of stability and biodegradability of chemical markers has not been investigated in sufficient detail.

1.3 Library-Independent Microbial Methods

One approach to MST is based on detection of host-specific microorganisms or host-specific ratios of organisms. A ratio of fecal coliforms to fecal streptococci of >4 has been considered indicative of human fecal pollution whereas a ratio ≤ 0.7 supposedly indicates a non-human source. However, variable survival rates, variable sensitivity to water treatment, differential persistence of enterococci relative to coliforms, the effects of animal diet on bacterial population structure, and the failure of this technique to discriminate between human and animal waste have resulted in a general abandoning of this technique. The ratio of atypical bacterial colonies to total coliform colonies on m-endo medium has also been suggested as an indicator of human fecal pollution (Booth and Brion, 2004), but the approach has not been widely investigated.

Potential species-specific indicators of fecal contamination include: *Streptococcus bovis* indicating non-human sources; *Bifidobacteria* spp. and the *Bacteroides fragilis* group signifying recent human contamination; *Clostridium perfringens* representing point-source sewage contamination; *Enterococcus faecalis* indicating human and/or bird contamination; and *Rhodococcus coprophilus*, which is considered specific for grazing farm animals and typically not found in human feces. *Bifidobacterium* spp. are early colonizers and a major component of the human fecal material but are rarely found in other animals. Therefore, they may provide an indication of human fecal contamination (Nebra et al., 2003). In addition, host-associated molecular markers in *Enterococcus faecium* have also been reported (Scott et al., 2005). A PCR-based assay targeting an *E. faecium* surface protein gene (*esp*) was positive with 97% (n = 65) of human sewage samples but negative with all animal fecal and waste lagoon samples (n = 102) representing poultry, pigs, cattle, and wild birds.

Enterococcus faecalis is frequently recovered from human, chicken, and seagull feces but not from cows, horse, pigs, rabbits, sheep, or deer. Since broiler fowl litter, which is the most likely source of environmental contamination from domestic birds, appears to select against *E. faecalis*, then the detection of the organism in water may indicate human and/or wild bird contamination (Kuntz et al., 2004).

Bacteriophages of *E. coli* and *B. fragilis* may also indicate the source of contamination. There are currently four recognized F-specific RNA coliphage types with varying levels of specificity (Table 1-2): two of them (II and III) were reported to be associated with human/domestic sewage, whereas types I and IV predominated in animal waste (Hsu et al., 1995). In one application of F-specific coliphage typing, the predominant source of contamination in a park was determined to be non-human (Griffin et al., 2000). Coliphages identified as groups I, II and III were identified in raw sewage samples, and groups I, III, and IV were identified in gull feces (Noble et al., 2003). In another study, all four types were identified in wastewater from pig farms, while types I, II, and IV were identified in cattle wastes, and types I, II, and III were detected in municipal wastewater treatment plants (Cole et al., 2003). Overall, the frequencies of F+ coliphage occurrence were higher in human- and cattle-affected waters compared to background waters. Detecting all phage types in municipal and farm wastes in these studies may be a reflection of multiple potential sources of input into wastewaters rather than a lack of specificity of coliphage types. The association of genotypes II/III with human waste, and I/IV with non-human animal feces was statistically significant but not absolute (Schaper et al., 2002b) indicating that coliphage typing is not a definitive source tracking tool.

Vinje et al. (2004) found groups II and III primarily in environments influenced by human waste and groups I and IV mostly associated with animal pollution. In addition, the four groups were found to be highly divergent based on replicase gene sequences with some sub-group divergence and a potential novel subgroup. Also, three genetic clusters were identified among F+ DNA coliphages (designated as M13, fd, and CH clusters) based on a 190 nucleotide region of gene IV. The authors suggested that the sub-grouping of coliphage types and additional groups may improve the applicability of coliphage typing for MST.

The presence of M13-type DNA coliphages in human wastewater and low level of occurrence in livestock waste suggests that this group of phage may also be a human-specific indicator (Long et al., 2005). Bacteriophages infecting human-specific *Bacteroides fragilis* strains have also been proposed (Puig et al., 1999) but the absence of these phages from some highly polluted waters and domestic sewage, and variability in their geographic distribution may limit their utility.

Table 1-2. Coliphage Types Detected in Fecal and Wastewater Samples.

Sample	Coliphage types	Reference
Human feces	II and III	Schaper et al., 2002b
Cattle feces, slurry, and abattoir waste	I and IV (89%), also II, III	Schaper et al., 2002b
Pig feces, slurry, and abattoir waste	I and IV (90%), also II, III	Schaper et al., 2002b
Poultry feces, slurry, and abattoir waste	I and IV (85%), also II, III	Schaper et al., 2002b
Untreated municipal sewage	II and III (75%), also I, IV	Schaper et al., 2002b
Treated sewage	II and III (73%), also I (57%), IV	Schaper et al., 2002b
Raw sewage	I, II, and III	Noble et al., 2003
Gull feces	I, III, and IV	Noble et al., 2003
Pig farm wastewater	I, II, III, and IV	Cole et al., 2003; Long et al., 2005
Cattle waste	I, II, and IV	Cole et al., 2003; Long et al., 2005
Wastewater treatment plant	I, II, and III	Cole et al., 2003
Birds	I and IV	Long et al., 2005
Human waste influence	II and III	Vinje et al., 2004
Animal waste influence	I and IV	Vinje et al., 2004

However, the coliphage typing approach may be limited by differential survival in water. For example, group I coliphages demonstrated greater persistence in natural waters compared to the other three groups and was the most resistant to a variety of stresses including exposure to ammonia, pH 10, elevated temperature (37°C), and salinity levels up to 42 g/L (Schaper et al., 2002a). In addition, group IV coliphages were significantly more sensitive to chlorination. Variable stability and sensitivity to environmental stress may lead to shifts in the distribution of phage groups over time, possibly biasing the results towards the more stable phages regardless of the actual source of contamination.

Culture-independent molecular methods such as PCR-based detection of 16S rRNA sequences from the *Bacteroides-Prevotella* group of bacteria or detection of *E. coli* toxin genes may also be indicators of host-specific contamination (Bernhard and Field, 2000; Bernhard et al., 2003; Oshiro and Olson, 1997). These techniques are rapid (usually requiring <24 hours), inexpensive, and do not require culturing of target microorganisms or site-specific libraries. PCR-based methods have also been proposed for human-specific *Bifidobacteria* spp. (Matsuki et al., 1998) and a multiplex PCR with strain-specific primer sets detected only *B. adolescentis* and *B. dentium* exclusively in human sewage samples (Bonjoch et al., 2004). However, these methods may not be applicable to all sources/types of water due to inhibitors that can be difficult to remove during DNA extraction. This may be particularly problematic in the most heavily polluted waters. In addition, the specificity of the primers may be broader than intended, depending on the number of non-target organisms that are tested. Also, toxigenic *E. coli* are infrequently isolated from the environment, even in the presence of high levels of contamination.

A 16S rDNA-based microarray targeting approximately 50 species of indicator bacteria that may have some degree of host-specificity (e.g. *Bacteroides* spp., *Bifidobacterium* spp., *Pseudomonas aeruginosa*, *Facklamia* spp.) was recently constructed and preliminary testing demonstrated the ability to detect human-specific indicators in wastewater samples (Berthiaume et al., 2004). However, considerably more optimization of amplification reactions, array hybridization, and selection of target organisms and sequences is necessary before this approach can be evaluated fully.

The main advantage of culture-independent molecular methods is their potential ability to rapidly identify the presence or absence of certain fecal sources in water, without requiring time consuming culturing or expensive library construction. However, as with all applications of molecular techniques to environmental samples, false-negatives and false-positives may be problematic. In a comparison of culture-independent, library-independent molecular methods, three causes of false-negatives and false-positives were reported (Table 1-3; Field et al., 2003). Overall, 41% of spiked samples failed to yield amplification products. Although PCR-based amplification of Bacteroidetes-specific 16S rDNA markers demonstrated the highest rate of successful amplification from spiked water samples (75%, N=24) and had a higher rate of correct source identification than most other methods, there was poor agreement between two laboratories conducting the same approach (Field et al., 2003). An overall limitation of these microbial indicator methods is their relatively course resolution; they may discriminate between human and non-human contamination but can not discriminate between different non-human animals. However, a significant advantage of the indicator methods is their relatively low cost that makes them available to water/wastewater utilities that do not generally need the higher levels of discriminatory resolution provided by more complex methods.

Table 1-3. Causes of Erroneous Results in Culture- and Library-Independent Methods.

False-negatives	False-positives
Operator error Insufficient sensitivity Interference by compounds such as humic acids in the sample matrix	Operator error (e.g., sample contamination) Insufficient specificity of the assay Insufficient specificity of the target genes or bacteria

The indicator methods may be limited by the rapid die-off of many organisms in water. Also, some of them are not exclusively species-specific so interpretation of results can be difficult. There has been little research conducted on the robustness of these methods or on the search for alternative species-specific microbial indicators. Some of the problems that still need to be addressed with these methods include:

- Replication of target organisms in the environment.
- Environmental reservoirs of target organisms.
- Persistence and survival in the environment, including variable survival rates.
- Variable sensitivity to water treatment processes.
- Variable and inconsistent occurrence of some indicators in polluted environments.

Human-specific viruses, such as some of the enteric viruses and human adenoviruses, as well as non-human viruses (e.g., bovine enteroviruses) may be useful in some circumstances. Such viruses may be detected directly in water using cell culture-based and/or molecular techniques. A variety of human enteroviruses has been found in recreational waters and waters that are sources of drinking water (Chapron et al., 2000; Schvoerer et al., 2001) indicating human sources of contamination. The apparently stringent host specificity of adenoviruses suggests their utility for MST although there are few studies on the prevalence of adenoviruses in water. Porcine and bovine adenoviruses were detected in 70% of pig samples and 75% of bovine fecal samples, respectively (Maluquar de Motes et al., 2004). In addition, bovine-specific enteroviruses may be indicators of bovine fecal contamination (Ley et al., 2002). A recent study using PCR-based detection reported human enteroviruses, human adenoviruses, and bovine enteroviruses in 57%, 37%, and 37% of water samples, respectively, from a river that receives sewage, industrial discharges, and run-off from urban development and agriculture (Fong et al., 2005). Teschoviruses that specifically infect pigs have also been proposed as indicators of porcine fecal contamination (Jimenez-Clavero et al., 2003). Other animal host-specific viruses such as Torque Teno Viruses (TTV), which are present in human and other animal populations and spread by fecal-oral transmission, may also be useful MST tools. Human-, non-human primate-, and porcine-specific TTV have been described (McKeown et al., 2004; Okamoto et al., 2000). However, relative isolation frequencies for enteric viruses in environmental waters tend to be low, although they are relatively common in sewage. The utility of these host-specific viruses has not been evaluated and the host-specificity of the various animal-specific viruses has not been fully assessed either. For example, deer, sheep, goats, horses and geese carry bovine enteroviruses (Jimenez-Clavero et al., 2005; Ley et al., 2002). In addition, current cell culture-based methodologies and analysis costs may limit the feasibility of this approach.

Source tracking has also been applied to individual pathogens based on PCR amplification of pathogen-specific DNA sequences that display host-specific polymorphisms. For example, there appears to be some host-adapted strains of *Giardia duodenalis*, possibly allowing differentiation between isolates from humans, livestock, and domestic animals (Rochelle, 2002). Recent advances in the molecular characterization of *Cryptosporidium* spp. suggest that there are host-specific species and genotypes within individual species. For example, *Cryptosporidium hominis* (previously referred to as genotype 1 *Cryptosporidium parvum*) has been isolated almost exclusively from humans, whereas *C. parvum* (previously genotype 2 *C. parvum*) occurs in a wide range of animals, including humans. Out of a total of 29 storm water samples analyzed by a PCR-RFLP analysis targeting the SSU rRNA gene, 93% were positive for *Cryptosporidium* with 12 different genotypes represented (Xiao et al., 2000). None of the detected genotypes matched those typically found in human, farm animals, or domestic animal samples. However, four were identical or closely related to *Cryptosporidum baileyi* (primarily associated with birds) and *Cryptosporidium* genotypes from opossums and snakes, indicating that wildlife was the primary source of oocyst contamination of surface water during storms. The predominant genotypes in untreated surface water matched the profiles of *C. hominis* (indicating human fecal contamination) and *C. parvum* while *Cryptosporidium andersoni* (prevalent in cattle) was most commonly detected in wastewater (Xiao et al., 2001). In addition, a compilation of genotyping data demonstrated that 15 of 22 waterborne outbreaks of cryptosporidiosis (68%) were caused by *C. hominis* (McLauchlin et al., 2000; Sulaiman et al., 1998). These data suggest that human contamination was responsible for the majority of outbreaks. However, targeting specific pathogens is only useful if the particular organisms are present in the body of water under investigation. The typically low frequency of occurrence of specific pathogens in natural waters limits the application of this approach for broad source tracking efforts. Also, as with the direct molecular techniques described above, these methods may not be applicable to all sample types because of the presence of recalcitrant PCR inhibitors in natural waters.

Although further evaluation of the various host-specific microbial methods, both culture-dependent and independent, is necessary to fully evaluate their specificity, temporal and spatial stability, robustness, and accuracy, they do not require the construction of large source libraries. Consequently, they are quicker and much cheaper than the library-dependent approaches.

1.4 Library-Dependent Phenotypic Methods

Phenotypic profiling, particularly antibiotic resistance analysis (ARA), has been most commonly used in small watersheds. However, their application may be less effective in larger geographical areas. The basic assumption underlying ARA is that exposure of humans and animals to a variety of antibiotic regimes (types, concentrations, frequencies), for either therapy or as food supplements, will result in the development of specific antibiotic resistance patterns in the intestinal microbial flora of the host animal. Exposure to different antibiotic regimes should therefore result in different, animal-specific resistance profiles. Based on this assumption, ARA targeting *E. coli* and enterococci has been proposed as an efficient method for MST (Hagedorn et al., 1999; Harwood et al., 2000; Wiggins et al., 1999) although there is considerable variation in the number, types, and concentrations of antibiotics used by various researchers, making cross study comparisons difficult. Since ARA is a library-dependent method, considerable research has focused on the ability of the technique to correctly classify animal isolates. A library of 1,174 known-source enterococci isolates analyzed by ARA had correct classification rates (CCR) of

88-95% for wildlife, livestock, and human isolates (Graves et al., 2002). When 2,012 stream isolates were compared to this library, approximately 10% could be attributed to a human source compared to 90% for wildlife and livestock. Wiggins et al. (2003) tested ARA-generated enterococci libraries of different sizes to determine how well they represented the target organism in various hosts. Smaller libraries had higher average rates of correct classification (ARCC) but were not particularly accurate at assigning the source of non-library isolates. Only the largest library tested (2,931 isolates) was able to classify non-library isolates with the same accuracy as library isolates. A merged multi-watershed library of 6,587 isolates had an ARCC of 57% (Wiggins et al., 2003). In the first application of ARA within the United Kingdom, ARCC values ranged from 69% for a six-way classification (municipal waste, cattle, pig, poultry, sheep, and wild animals) to 86% when differentiating between enterococci from municipal wastewater versus all other isolates (Ebdon et al., 2004). The best classification of wastewater isolates (72%, N = 403) was achieved using a total of 14 antibiotics, most at a single concentration, and wastewater isolates had the highest levels of resistance to antibiotics that are used exclusively for human therapy. Comparison of 8,542 river isolates with 1,451 known source enterococci isolates (RCC = 82% for human isolates) demonstrated 25%, 29%, and 48% contributions from human, wildlife, and livestock sources, respectively (Booth et al., 2003). This led to efforts to reduce and control contamination from all three types of source as part of the TMDL compliance strategy for this river.

Potential problems with the ARA approach include non-domesticated animals that are not exposed to antibiotics on a routine basis and whose microbial flora may have limited resistance to the antibiotics used. Such zero-resistance isolates can bias source assignment results and would usually need to be excluded from the analysis. Antibiotic resistance determinants are frequently carried on mobile genetic elements such as plasmids that can be lost from cells or transferred between cells in response to selective pressures. Plasmid transfer occurs within animal intestines and in natural aquatic environments. In addition, because of differential selective pressures, resistance profiles may vary across geographical regions and time. However, at least one study has demonstrated the stability of antibiotic resistance patterns for source assignment over a one-year period (Wiggins et al., 2003).

Carbohydrate utilization profiles (CUP) of *E. coli* and enterococci has also been suggested as an MST method (Hagedorn et al., 2003; Wallis and Taylor, 2003) and performed equally to ARA methods in a recent comparison (Harwood et al., 2003). Utilizing 30 of the 95 substrate wells available in BIOLOG plates, the ARCC for 365 known source fecal enterococci isolates were 93% for a human versus non-human classification and 86% for a three-way classification (Hagedorn et al., 2003). Ninety percent of stream isolates from sites suspected to be influenced by human contamination were classified as human. However, few evaluations or extensive testing of the discriminatory resolution or reproducibility of the method have been performed. In a comparison of ARA targeting *E. coli*, enterococci, fecal streptococci, and fecal coliforms, along with CUP of *E. coli* and fecal streptococci, the ARCC ranged from approximately 51-85% (Harwood et al., 2003). In most instances, the rate of false-positive identification of isolates was as high as the correct positive rate. Based on the predictive accuracy for water samples, ARA targeting fecal streptococci was the best method (true-positive rate of 100% compared to false-positive rate of 39%), although only 12 water samples were analyzed so this result may not be supported statistically. Further conclusions of this study were that the size of the libraries (~300 isolates) was not large enough to avoid overlap between source assignments and that the internal accuracy of the libraries (based on ARCC) did not correlate with the accuracy of source prediction in water samples. It is often stated that phenotypic methods are relatively inexpensive compared to genotypic methods, allowing larger libraries to be constructed and more water isolates to be analyzed. However, this statement may not be accurate when considering the amount of labor involved in the various methods, labor pay rates, and the relative cost per sample of high throughput molecular methods. In addition, it is not clear that true cost comparisons have been made between methods.

1.5 Library-Dependent Genotypic Methods

Molecular fingerprinting tools such as pulsed field gel electrophoresis (PFGE), ribotyping, and various PCR-based methods have only been applied to MST of environmental fecal contamination relatively recently. However, many of the same techniques have been used for over 15 years to trace clinical and foodborne isolates of bacteria during disease outbreaks and to track endemic occurrence of diseases (Swaminathan et al., 2001). As an example, PulseNet is a PFGE-based surveillance program coordinated by the Centers for Disease Control and Prevention (Swaminathan et al., 2001). Fifty state public health laboratories and numerous other facilities submit PFGE data on isolates of *E. coli* O157:H7, *Salmonella* spp., *Shigella* spp., and *Listeria monocytogenes* to a national database. Because all of the laboratories follow a standardized protocol, PFGE patterns can be compared within the network of laboratories allowing associations to be made among isolates, greatly aiding in the identification, surveillance and management of outbreaks. While recognizing that environmental microbial communities are typically far more diverse than clinical communities, the PulseNet program serves as an example of the successful application of a molecular typing technique to investigate disease epidemiology. Microbial source tracking in water is similar to clinical epidemiology and may be considered as a type of environmental epidemiology.

In PFGE the entire genome of the target organism is digested by a restriction enzyme with a low cutting frequency (typical PFGE profiles contain 15 – 25 bands) and separated by gel electrophoresis (Figure 1-1). Even with highly standardized protocols, there is something of an art to reproducible PFGE analysis. For example, small fluctuations in experimental conditions can decrease each fingerprints comparative value. Nevertheless, PFGE analysis is considered the "gold standard" for epidemiological typing of bacteria. The method is highly discriminatory but may not be necessary for source tracking in an environmental context (McLellan et al., 2003) because it results in too many unique banding patterns and difficulty in assigning sources. In a comparison of seven MST protocols, PFGE was the most accurate, correctly identifying 100% of 24 isolates in a subset of library isolates (Stoeckel et al., 2004). The ARCC for PFGE was 41% for classification of isolates at the host species level, 72% when differentiating between humans, domestic animals and wildlife, and 79% for a human versus non-human classification. These values were higher than for any of the other methods (except ribotyping with *Eco*RI, but less than 6% of isolates were tested by this technique). Conversely, Parveen et al., (2001) reported that there was no association between PFGE profiles and the source of the isolate, although this cannot be discerned from the PFGE results presented in the paper.

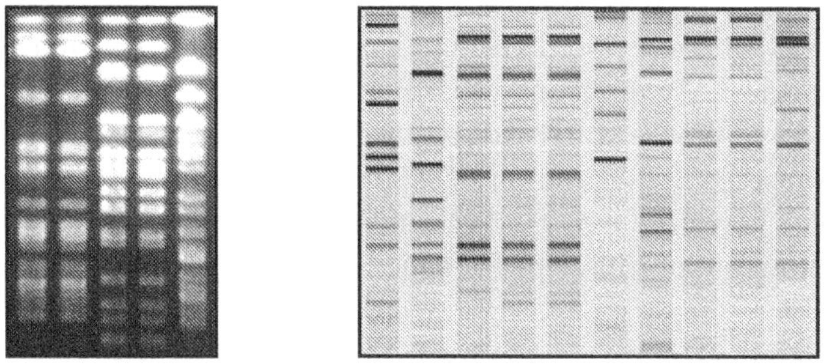

Figure 1-1. Typical Pulse-Field Gel Electrophoresis (left panel) and REP-PCR Profiles (right panel).
Results obtained with environmental isolates of *Escherichia coli* (Rochelle, Hacker, and De Leon, unpublished data).

Repetitive element PCR uses primers that target short repetitive DNA sequences distributed throughout an organism's genome to generate a banding profile typically containing 15 to 40 bands (Figure 1-1). These repetitive DNA elements occur in most bacteria (Hulton et al., 1991; Martin et al., 1992; Stern et al., 1984; Versalovic et al., 1991) and the techniques are known collectively as repetitive (rep)-PCR. Although they are all based on the same principle, the rep-PCR techniques vary in their discriminatory power and reproducibility due in part to the frequency and uniqueness of the target repeat elements. Primers target:

- Repetitive extragenic palindromic sequences (REP-PCR).
- Enterobacterial repetitive intergenic consensus sequences (ERIC-PCR).
- A third class of extragenic repeating elements (BOX-PCR).

Dombek et al. (2000) reported that BOX-PCR (ARCC = 93%) was more effective than REP-PCR (ARCC = 66%) for grouping 208 *E. coli* isolates from various animal hosts. Although multivariate analysis of variance of BOX-PCR fingerprints resolved all isolates into discrete host-specific clusters, cluster analysis using the Jaccard similarity coefficient and nearest neighbor tree construction displayed considerable overlap between isolates from various hosts. In a comparison of rep-PCR using the BOXA1R primer and ribotyping of 482 *E. coli* isolates from humans and a variety of animals, BOX-PCR had better reproducibility and was more accurate in assigning isolates to their respective host animals (Carson et al., 2003). The ARCC across all animal types was 79% for ribotyping compared to 88% for BOX-PCR although there was no significant difference between the two methods in individual rates of correct classification (RCC) for four of the eight host species.

Analysis of 2,466 *E. coli* isolates from humans and 12 other animal sources by BOX-PCR and a derivative method using a fluorescently labeled BOX primer demonstrated an ARCC of 82% (Johnson et al., 2004). The authors concluded that removal of duplicate genotypes (isolates with identical fingerprints) within libraries, increased library size, appropriate statistical analysis methods, and correct alignment of band data within and between electrophoresis gels are all necessary for improved accuracy of MST methods. Removal of isolates with duplicate banding patterns left 1,535 unique isolates and reduced the ARCC to 60.5%, indicating a potential bias introduced when libraries contain multiple identical banding patterns. The authors suggested that the clonal nature of *E. coli*, which can result in multiple identical isolates originating from individual source animals, may bias the ARCC, alters database fidelity, and overestimates the ability of a method to correctly assign isolates to source animals (Johnson et al., 2004). Based on REP-PCR, 57% of *E. coli* isolates from beach sand and lake water were unique with fingerprint patterns that did not match any other isolate (Kinzelman et al., 2004). In addition, there was no correlation between similarity scores of pairwise comparisons for REP-PCR and cellular fatty acid analysis. In some instances, the authors reported that isolates with almost identical REP-PCR patterns had different cellular fatty acid profiles. Similar poor correlations between techniques were reported by McClellan et al. (2003) who studied 440 isolates of *E. coli* from humans, dogs, gulls, cattle, and sewage treatment plants. Only 40% of isolates that were identical by REP-PCR were also identical by ERIC-PCR and PFGE. In some instances, strains that were indistinguishable by REP-PCR and ERIC-PCR were differentiated by PFGE.

Amplification of the 16S-23S rRNA intergenic spacer region (ISR-PCR) was compared to BOX-PCR for source assignment of *E. coli* isolates (Seurinck et al., 2003). Based on similarity thresholds of 68% for ISR-PCR and 49% for BOX-PCR, 87 (ISR-PCR) and 59 (BOX-PCR) distinctive fingerprints were generated from 267 isolates. The authors also reported that while most of the animal samples generated multiple banding patterns, one unique horse-specific fingerprint was detected by BOX-PCR. The ARCC was 84% for ISR-PCR and BOX-PCR combined. These authors also used the Chao1 estimate of population richness (Chao, 1984) to determine that the entire *E. coli* population of particular sources was represented by 15 isolates for raw sewage, 22-25 isolates for horse samples, and 32 for gulls (Seurinck et al., 2003). Random amplification of polymorphic DNA (RAPD) with primers generating 2-16 amplification bands was also reported to be useful for discriminating between *E. coli* isolates from human and other animals (Venieri et al., 2004) although the lack of profile variability from human isolates suggests that the authors were analyzing many clonal isolates.

The reproducibility of molecular fingerprinting methods can be easily evaluated by multiple analyses of a single isolate. Based on the Dice coefficient of similarity and clustering by an unweighted pair group averaging method, Seurinck et al. (2003) used the similarity value of the cluster that contained replicate rep-PCR analyzes of a single isolate as the similarity cutoff. The resulting similarity cutoff was 49% for BOX-PCR and 68% for ISR-PCR. These figures indicate a lack of reproducibility in the fingerprinting or analysis methods. The reproducibility of rep-PCR using ERIC2 and BOX1AR primers was increased by modifying amplification conditions (primarily elevated annealing temperatures) but rep-PCR could not resolve all phylogenetic MLEE-based groupings (Johnson and O'Bryan, 2000).

Ribotyping is widely used for identification and characterization of diverse bacteria in clinical laboratories because it is one of the most highly standardized and reproducible DNA banding based methods and the process has been fully automated. In the first application of ribotyping to MST, Parveen et al. (1999) analyzed 238 *E. coli* isolates with four enzymes and reported that digestion with HindIII provided the best discrimination between human and non-human fecal isolates with an ARCC of 82%. However, at a similarity index of 50%, all isolates were grouped into four general clusters, each of which contained isolates from both human and non-human sources. Eight ribotype profiles were unique to human isolates and another eight were unique to non-human sources. The remaining profiles were generated by both human and non-human isolates. In contrast to the above study, the results of a recent methods comparison demonstrated that ribotyping with EcoRI was more accurate than with *Hind*III (Stoeckel et al., 2004), although only nine isolates were analyzed by EcoRI compared to 150 for HindIII making comparisons difficult. The ARCC values for HindIII ribotyping in this study were 13% for an eight-way classification, 33% for a three-way, and 49% for a two-way human vs. non-human classification. In the overall comparison of six library-dependent methods (four genotypic and two phenotypic), HindIII ribotyping consistently achieved a correct classification rate equal to or less than that expected by random chance. In another study, HindIII ribotyping could not differentiate between *E. coli* isolates from beef cattle, dairy cows, poultry or swine with an overall misclassification rate of 65.3% (Scott et al., 2003). However, 84.5% of human isolates were correctly identified (N=84) and 77% of animal isolates (N=317) were identified as non-human, leading the authors to conclude that it may be possible to differentiate human from animal-derived *E. coli* over relatively broad geographic regions using ribotyping. Ribotyping *E. coli* isolates using EcoRI and PvuII was reported to differentiate between cattle, horse, pig, and chicken isolates, although there were common ribotype patterns in different host animals (Hartel et al., 2002). Also, geographic variability precluded the universal use of a library developed from a limited geographic region.

Denaturing gradient gel electrophoresis (DGGE) of PCR-amplified *uidA* sequences was proposed as a source tracking tool and demonstrated 11 sequence types within a collection of 50 freshwater isolates and reference strains (Farnleitner et al., 2000). However, it has not yet been applied to isolates from a variety of hosts and sources. The *uidA* gene was also the target of sequence-based differentiation of *E. coli* isolates (Ram et al., 2004). A total of 941 isolates exhibited 114 different alleles represented by up to 17 nucleotide differences. Out of 81 alleles identified in isolates from various animal fecal samples, some were observed exclusively in particular animals but many occurred in several host species; the rate of correct classification was 60%.

In a comparison of genotypic methods (rep-PCR, ribotyping, and PFGE) mostly targeting *E. coli* and involving six investigators, none of the methods stood out as superior to the others for correctly identifying fecal sources (Myoda et al., 2003). All of the methods were able to identify the dominant source in most of the blind samples but the methods also had high false-positive rates (up to 57% for ribotyping by one investigator). The majority of the differences in method performance were investigator-dependent rather than method-dependent, indicating the crucial necessity of optimization and standardization.

1.6 Data Analysis

Library-based methods in particular can generate a lot of complex data and the methods used to analyze these data become a critical component of the overall source tracking approach. As data sets become larger, more sophisticated analysis methods are required to tease out patterns of classification. Statistical methods that have been applied to MST-related studies include discriminant analysis, multivariate analysis of variance, maximum similarity, average similarity, nearest neighbor, principal component analysis, multiple regression, logistic regression, and classification and regression tree analysis. Ritter et al. (2003) compared seven statistical methods to analyze data from three rep-PCR and three ARA libraries. The ability of the methods to correctly assign isolates to their respective sources was highly dependent on which statistical method was used. Within a single library generated by one method, correct classification scores ranged from 0-100% depending on the statistical method used and source animal. Also, across all six libraries correct classification scores ranged from 0-100% for a single analysis method. In addition, changing parameters within a single statistical method, such as pooled covariance versus non-pooled covariance for discriminant analysis, or using different methods for estimating distance within a nearest neighbor analysis, could dramatically alter the outcome of source assignment. Overall, no single statistical method proved superior across all libraries (Ritter et al., 2003). Such inconsistency may be disconcerting. Using the example of phylogenetic analysis of DNA sequence information where distance and similarity calculation methods are used to infer phylogenetic/evolutionary relationships between sequences and their host organisms, one approach to assessing the robustness of phylogenetic analyses is the use of multiple analysis methods to calculate distance matrices and infer phylogeny. Congruence by multiple methods increases confidence and implies that the correct relationships have been inferred. Conversely, a lack of agreement between different analysis methods decreases confidence and may question the underlying assumptions of the approach or the quality of the sequence data. It has been suggested that uncertain or poorly matched isolates should be removed from MST analyses. Again using the phylogenetic analysis of DNA sequences as an example, it is generally recognized that distance matrices should be calculated based only on aligned sequences and that ambiguously aligned sequences and unaligned nucleotides are removed because of the bias they introduce; the phylogenetic tree may reflect analysis bias rather than true evolutionary relationships.

The accuracy of genotypic methods that produce banding patterns is likely to be heavily influenced by the similarity cut-off that is selected. Some studies have considered a similarity score of at least 75% as indicative of related isolates while others only consider a match if banding patterns are identical. There is little consensus on the similarity score above which isolates are considered to match (similarity thresholds). Lower threshold scores result in easier classification of an isolate to a particular source but the accuracy of such classification may be circumspect. At the other extreme, only permitting exact matches increases the confidence in source apportionment but may reduce the likelihood of being able to identify a source. There is a direct correlation between the similarity threshold selected in a particular study and the resolution of clustering, regardless of the number of isolates examined (Figure 1-2).

Library-based methods typically use binary band matching algorithms such as Jaccard or Dice coefficients of similarity to estimate the similarity between isolates. Alternatively, molecular fingerprints can be represented as densitometric curves (which consider band position and intensity) that are analyzed using Pearson's product-moment correlation coefficient (Hane et al., 1993; Johnson et al., 2004). This approach to estimating fingerprint similarity resulted in an ARCC of 60% for a library of 1,535 *E. coli* isolates representing 11 animal sources compared to 44% when using the Jaccard band matching method (Johnson et al., 2004).

Since MST methods do not exclusively assign isolates to an individual source, one problem in the interpretation of MST data is the threshold at which identified source identifications are considered reliable. Some researchers have calculated a minimum detectable percentage (MDP) for particular methods. An MDP of 25% means that if at least 25% of watershed isolates are assigned to a particular source, it can be assumed that this is a true source (Wiggins et al., 2003). If less than 25% of isolates are assigned to a particular source they may represent a misclassification of other source types. However, this approach does not take into account true sources that account for less than the MDP of the isolates in a watershed.

ARCC is a frequent measure of the accuracy of individual source tracking methods. However, ARCC is dependent on the size of the library analyzed (Figure 1-3) up to a point and eventually reaching an asymptote. Smaller libraries will tend to have higher ARCC values but are less representative than larger libraries. Therefore, small libraries may not be useful when using ARCC values to assess the ability of a method to assign isolates to their respective host/source animals. In many studies with both genotypic and phenotypic library-based methods, ARCC values typically range from approximately 50–95% and trend towards decreased ARCC as library size increases. However, in a recent evaluation of MST methods (Stoeckel et al., 2004), ARCC values were considerably lower than those reported by most previous studies (Figure 1-3).

Construction of a library (also called a host origin database to avoid confusion with clone "libraries") requires analysis of isolates from all of the potential source animals. The size of the library seriously affects interpretation of source tracking data. Increasing library size will increase its representation of potential source animals and increase the statistical validity of the results. Conversely, larger libraries will have reduced ARCC values leading to lower confidence in the methods by end users. In addition, one library is not likely to be applicable to studies covering wide geographic areas. It may be necessary to develop a separate library for each geographic area, watershed, lake, or stream under investigation (Hartel et al., 2002).

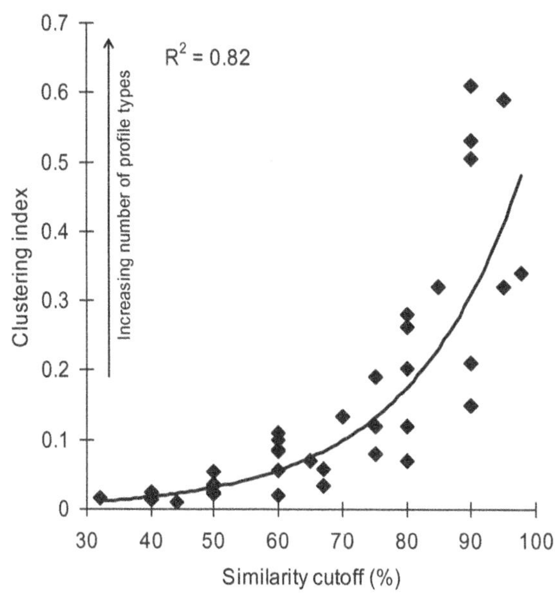

Figure 1-2. Relationship Between Similarity Threshold and the Resolution of Cluster Analysis Methods.
The clustering index was calculated as the number of distinct clusters at each similarity cut-off divided by the total number of isolates analyzed. Data were extracted from studies that used ribotyping, REP-PCR, BOX-PCR, RAPD, DGGE, and PFGE (Avery et al., 2002; Dombek et al., 2000; Farnleitner et al., 2000; Jenkins et al., 2003; Johnson et al., 2004; McLellan et al., 2003; Pacheco et al., 1997; Parveen et al., 1999; Venieri et al., 2004; Wheeler et al., 2002; Yang et al., 2004).

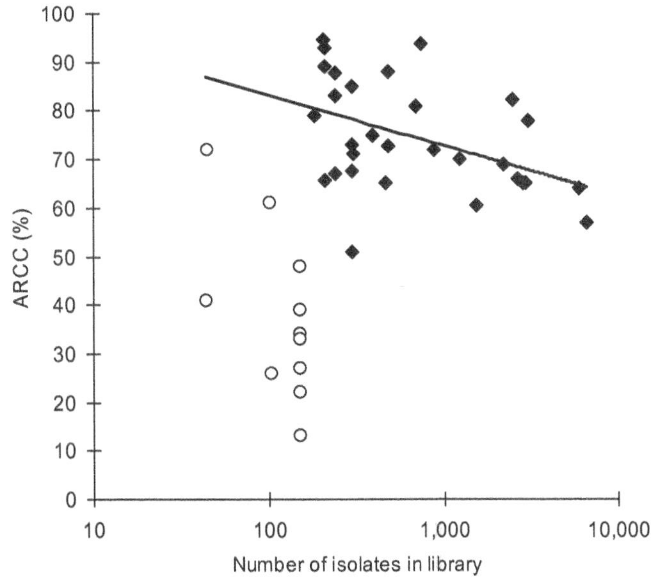

Figure 1-3. Relationship Between Library Size and the Average Rate of Correct Classification (ARCC) for MST Methods. Data (♦) were extracted from studies with both phenotypic and genotypic methods (Carson et al., 2003; Dombek et al., 2000; Ebdon et al., 2004; Graves et al., 2002; Harwood et al., 2003; Johnson et al., 2004; Wiggins et al., 1999; Wiggins et al., 2003). Data from a recent methods comparison were also plotted but not included in the regression analysis (○, Stoeckel et al., 2004).

1.7 Overarching Issues

The types of molecular technologies used for source tracking have been used successfully in other fields. For example, because humans are not clonal (i.e., the DNA of every human is unique), this uniqueness affects the type of DNA typing used in forensic investigations. In contrast, because bacteria divide by binary fission, many subspecies of bacteria are clonal. The PulseNet program provides an example of the successful application of a molecular technique (PFGE) to tracking the origin of these clonal bacterial strains. For example, in foodborne outbreaks, there are a limited number of possible sources, the bacterial contaminant is identified prior to the tracking efforts, and the contamination occurs in a medium (food) that has a small natural microbial population. Unfortunately, MST in natural waters presents the worst combination of factors:

- Potentially multiple locations at which contamination is introduced.
- Many potential animal sources.
- Clonal target organisms.
- A potentially vast background of naturally occurring microbial populations.

The general conclusion from the majority of MST studies examining the utility of the methods appears to be that more study is necessary to a) improve the methods, b) compare techniques, and c) gain stronger insight into the interpretation of the results. However, continually investigating all of the potential methods as well as newly developed methods stretches resources too thin. The concept of MST is based upon the assumption that there are stable, host-adapted species or strains within a species that are phenotypically or genotypically distinct from other species or host-adapted strains of the same species. As stated by Gordon

(2001), regardless of the technique or target organism, the successful application of microbial source tracking methods depends on the validity of several assumptions:

- Clonal composition of microbial populations differs among geographical localities.
- Species exhibit host specificity.
- Clonal composition of representatives of the species isolated from the environment represents the clonal composition of the species in the host animals responsible for fecal input.
- Clonal composition of bacterial populations is stable through time.

Therefore, a key issue is the validity of MST's underlying principle of unique host-adapted species/strains. If this assumption is invalid or valid only in some instances, then no amount of refining and comparing techniques, or developing new methods will improve reliability of MST tools to the point where it can be categorically stated that a particular source is responsible for any given contamination event. If the assumption is valid then there is a need to focus on a handful of the most appropriate and most promising methods. A multi-national European workshop selected the following methods as the as most promising and deserving of further investigation for source tracking in coastal waters (Pond et al., 2004; Rangdale, 2003):

- Adenovirus detection
- Genotyping *Bacteroides* spp.
- Detection of sorbitol-fermenting *Bifidobacteria* spp.
- Coliphage genotyping
- Enterotoxin biomarkers
- Fluorescent whitening agents
- Multilocus sequence typing (MLST)
- Pulse field gel electrophoresis (PFGE)
- Random amplification of polymorphic DNA (RAPD)
- Repetitive-PCR (rep-PCR)
- Restriction fragment length polymorphism (RFLP)
- Vibrational spectroscopy

Although some of these technologies are relatively mature and have proven useful in selected MST studies, the application of others to MST is yet to be assessed. The European workshop also suggested that the most suitable method(s) would target a source-specific marker(s) that meets the following criteria:

- Not dependent of a library that is susceptible to biological variability.
- Cost effective.
- Sufficiently characterized to ensure source specificity.
- Temporal and spatial stability assessed through extensive validation.
- Established relationships with compliance parameters.
- Should not be easily degraded in the marine or estuarine environment.
- Shed by a majority of the target population.
- Should be quantifiable and amenable to standardization.

In summarizing an earlier MST workshop, Stewart et al. (2003) stated that while library-dependent methods are currently the most frequently used, library-independent methods are likely to gain a stronger focus in the future, and recommended that MST concentrate on:

- Selecting better restriction enzymes and antibiotics for library-based methods.
- Defining the most appropriate library size.
- Selecting better target organisms.
- Selecting pattern-matching algorithms.
- Identifying new genomic targets.
- Developing quantification procedures for library-independent methods.
- Integrating microarrays and other direct pathogen detection methods into MST approaches.

Recent large-scale multi-laboratory evaluations of MST methods (Griffith et al., 2003; Harwood et al., 2003; Moore et al., 2005; Myoda et al., 2003; Stewart et al., 2003; Stoeckel et al., 2004) have cast some doubt on the potential of MST methods to identify individual sources of contamination. Following a comparison of seven library-dependent MST protocols (ARA, CUP, ribotyping with two enzymes, PFGE, BOX-PCR, and REP-PCR) with a 900-isolate library of *E. coli*, the authors concluded that better representation in known-source libraries and improved classification accuracy are needed before the techniques are reliable for field application (Stoeckel et al., 2004). The high rate of false-positive classifications was seen as an impediment to data interpretation for most protocols. When attempting to assign isolates to eight source animals, ARCC values ranged from 13% for CUP and ribotyping to 41% for PFGE. Considering the random expectation for eight source animals would be an ARCC of 12%, values of 13% do not inspire confidence. The ARCC values were higher when considering just a two-way classification into human or non-human sources (49–79%), but still some methods performed worse or only slightly better than would be expected by chance (ribotyping, ARA, and CUP). ARA, CUP, ribotyping with HindIII, BOX-PCR, and REP-PCR would have made inaccurate reports of contributing sources. PFGE would have reported contributing sources with reasonable accuracy but the interpretation would be hampered by the high rate of "unknown" classifications.

An MST workshop conducted in 2002 listed the following questions that needed to be addressed by method evaluation studies (Griffith et al., 2003):

- Can an individual laboratory produce repeatable results in multiple runs of the same sample?
- Can investigators in different laboratories produce consistent results for the same samples?
- Can methods accurately identify mixed bacterial sources in laboratory-created aqueous matrix samples?
- Can methods accurately detect the primary source(s) in samples from dominant-source watersheds?
- Do different methods produce comparable results in samples from complex natural systems that contain multiple unknown sources?

A series of method comparisons and evaluations were designed based on these criteria. The techniques evaluated were ARA, CSU, ribotyping, PFGE, BOX-PCR, Bacteroidetes-specific PCR, toxin gene-specific PCR, tRFLP, coliphage typing, and detection of human-specific enteroviruses and adenoviruses (Griffith et al., 2003). Most methods correctly identified spiked samples that contained human fecal material but false-positive rates were high (>80% for ARA and CSU). False-negative rates were also high (up to 100%). Methods targeting human viruses or coliphage identified less than 30% of samples that contained fecal material from individual humans but had a higher success rate with samples that contained raw sewage (40–90%). Although the virus methods did not identify most of the individual human samples as human sources, the methods themselves may not have failed. Rather, it may reflect on a failure of the assumptions underlying the approach (e.g., the individual human fecal samples may not have contained the target organisms if they were collected from healthy individuals). The methods with the highest rates of correct identification were PFGE and Bacteroidetes-specific PCR.

Moore et al. (2005) evaluated *Hind*III ribotyping and ARA of *E. coli* (997 isolates) and ARA of entereococci (3,657 isolates) with blinded proficiency samples. Animal source prediction accuracy for ARA was 28% for *E. coli* and 45% for enterococci. Ribotyping correctly classified 27% of *E. coli* isolates. The authors concluded that neither method is currently suitable for application to environmental samples.

An additional potential problem with studies designed to evaluate methods is identifying the target organism. "Known" organisms from potential sources are used to test the classification accuracy of MST methods. But what measures are taken to ensure that these organisms are correctly identified in the first place? For example, plating fecal samples on an *E. coli* selective media does not guarantee that all of the colonies that grow will be *E. coli*. If some of these false-*E. coli* are used to assess the classification accuracy of a particular method, it should not be surprising that poor results are obtained.

Many source tracking methods require culturing the particular target organism prior to profiling. However, culture-dependent methods can introduce their own biases. Many species are not readily cultured from environmental samples and even within a single species there will be differential cultivability. Most of the molecular typing techniques (PFGE, ribotyping, rep-PCR, etc.) target *E. coli* because it has traditionally been used as an indicator of fecal contamination in water. However, Gordon (2001) concluded that geographical location or host specificity accounted for little of the observed genetic diversity in *E. coli*. Even extreme inter-continental geographic separation accounted for only 10% of the diversity in *E. coli* populations (Souza et al., 1999). In one study, the taxonomic grouping of host animals explained only 6% of the observed genetic diversity among *E. coli* clones (Gordon and Lee, 1999). There is also considerable temporal variation in *E. coli* populations. According to one study that used multi-locus enzyme electrophoresis (MLEE) and detection of virulence factors, approximately 25% of within host variation in clonal diversity was due to temporal variation (Gordon et al., 2002). A second study found 240 ribotypes among 451 isolates from two herds of yearling steers and reported that the majority of ribotypes were transient and unique to each sampling time and animal (Jenkins et al., 2003). The authors suggested that it may be necessary to ribotype >900 *E. coli* isolates per host animal to overcome the limitation imposed by temporal variability. Furthermore, a large number of *E. coli* strains can be isolated from host animals within a species living in close proximity and the clonal diversity of *E. coli* changes as a result of adapting to environmental conditions (Gordon, 2001; Gordon et al., 2002). There appears to be substantial changes in the community structure of *E. coli* populations during the transition from the host gut to the environment. There are strains that are better adapted to conditions in the external environment compared to strains isolated from the gut (Gordon et al., 2002). Also, strains of *E. coli* in an animal gut that are best suited to survival in the environment may not be the dominant strains in the gut and so are likely to be missed by profiling methods applied to the host animal. These issues of *E. coli* population structure have led one author to suggest that assumptions implicit in MST methods that depend on commensal isolates of *E. coli* are probably invalid (Gordon, 2001). This conclusion may have been validated by a recent study in which the majority of seven library-dependent methods performed relatively poorly (Stoeckel et al., 2004). Other enteric bacteria such as *Citrobacter freundii* or *Hafnia alvei* may be more appropriate for tracing the source of contamination since they exhibit greater host specificity and may have increased spatial and temporal stability (Gordon, 2001).

The other primary target for library-based methods is enterococci. *Enterococcus faecalis* was identified in fecal samples from humans, dogs, and chickens but not from cattle, deer, pigs, or geese (Wheeler et al., 2002). The authors suggested that the potentially limited host range of *Ent. faecalis* combined with ribotyping may be a useful MST method. At a 90% similarity index, there was good separation between human and chicken isolates, but in many discrete sub-clusters rather than exclusive host-specific clusters. Also, no enterococci were recovered from two human fecal samples highlighting a potential limitation of this target organism. In addition, host-associated molecular markers in *Ent. faecium* have also been reported (Scott et al., 2005). As with *E. coli*, geographic and temporal variability may also be an issue with *Ent. faecalis* and *Ent. faecium* but there have been insufficient studies to make this assessment. Important additional considerations in the selection of target organisms are the survival and regrowth of indicator bacteria in water and sediments and the potential role of sediments as a reservoir for fecal indicator bacteria.

When methods are applied to actual case studies, investigators often conclude that there were insufficient isolates in the library. Some investigators have concluded that libraries should contain many thousands of isolates. Increasing the size of the library can dramatically increase a project's budget and unfortunately, available resources have to be one of the considerations in assessing the feasibility of source tracking programs. Utility managers are likely to ask how many isolates would be necessary to guarantee certainty. The size and composition of the source library are critical factors influencing the accuracy and reliability of library-based MST methods. Libraries must be representative of the sources in a study area and representative of the population diversity within individual source species. The approach used to construct a library can make the difference between a source tracking study that generates useful results and one that simply reports that there are many different sub-types of a particular organism in water. Important questions that have to be considered for library-based methods are:

- How to ensure the adequate representation of all sources within the library?
- How to ensure libraries will allow correct classification of unknown isolates?
- What is a statistically appropriate library size?
- How many different host species need to be included in the library?
- How many samples should be collected from each type of source animal?
- How many isolates should be collected from individual animals?

As the size of libraries increases, it may be necessary to automate techniques through robotics to increase throughput and reduce manual labor costs. Although analysis of an individual isolate is not particularly expensive, increasing library sizes can make library-dependent methods cost prohibitive. The minimum number of isolates needed to perform statistically sound studies has not been determined and there is still considerable debate on the applicability of universal libraries. It is likely that a new library will need to be constructed for each project location or watershed. The cost-benefit ratio of increasing library size must be a consideration in selecting an MST strategy since a large increase in library size may provide only a small gain in the confidence of the results but may result in an exponential increase in project costs. At what point is a study not cost-effective and at what size library is the threshold of diminishing returns reached?

An important consideration in the decision to use an MST approach, and in which method(s) to use is the required level of resolution (Figure 1-4). Three potential levels of discrimination are available:

- Human vs. non-human (two-way discrimination).
- Human vs. agricultural animals vs. wildlife (three-way discrimination).
- Human vs. cows vs. horses vs. sheep vs. pigs vs. dogs vs. seagulls vs. geese, etc. (eight-way discrimination).

Within these broad classification classes there will also be varying degrees of overlap between animals or host groups. It is likely that most water/wastewater quality managers would be primarily interested in differentiating between human and animal contamination. Discriminating between different animal sources is likely to be secondary objective or pursued as special studies. Also, recognizing that many of the methods generate false-positive and false-negative results, is there an acceptable level of ambiguity in source assignment? As described above, some of the current methods are reasonably robust at providing the first level of discrimination (human vs. non-human) and so selected MST methods provide a viable option. More optimization, refinement, and analysis will be necessary before a high degree of confidence can be placed on greater levels of resolution.

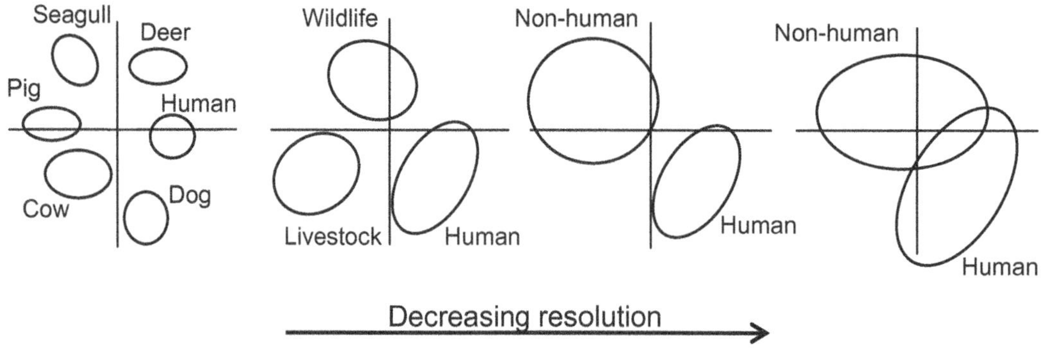

Figure 1-4. Theoretical Scenarios Describing the Levels of Discriminatory Resolution that May be Possible with Source Tracking Technologies.

1.8 Summary

MST technologies offer the potential to identify source(s) of fecal contamination in water. Such information is critically important to the wastewater, storm water, and drinking water industries. However, there are many proposed methods encompassing chemical and biological targets, phenotypic and genotypic techniques, culture-dependent and independent methods, and library-dependent and independent methods. In general, MST methods are not well established and have not been used extensively in “real world” situations. There have been attempts to standardize and validate some protocols, but these protocols are not widely available. Their adoption by non-specialist laboratories would require substantial training and resources. In addition, issues of quality assurance have not been addressed in sufficient detail. Decisions on the applicability and acceptability of particular methods should include a discussion of the quality assurance practices. In concluding their review of MST methods, Simpson et al. (2002) stated that none of the currently used techniques has been systematically applied to evaluate the population dynamics of the targeted organisms on a spatial or temporal basis. Research is needed on the effect of environmental and biological factors, within the primary habitat (animal gut) and secondary habitats (environmental waters, watersheds), on the stability of an organism’s genotypic or phenotypic profile. Nevertheless, based on the recent relatively rigorous comparative studies that have been performed, some MST methods may be able to discriminate between human and non-human sources of contamination.

CHAPTER 2.0

WORKSHOP OVERVIEW

The water and wastewater industries are interested in developing a better understanding of the sources of fecal contamination. By definition, the field of MST offers the potential to apportion the contribution of various animal groups to this contamination. However, while there are many methods available there is no clear indication as to the most appropriate approach. There has been little systematic comparison of methods and only a few blind trials, and issues of reproducibility, accuracy, sensitivity, specificity, resolution, and robustness need to be addressed. In 2002 a workshop on MST was organized by the Southern California Coastal Waters Research Project against a background of confusion among MST customers, confusion among some scientists, and the search for a "silver bullet" technique. The 2002 workshop helped to reduce the confusion and enhanced communication among researchers. However, the resulting method evaluation studies served as a reality check for MST researchers and were a source of consternation for many potential end-users. Three years on there is still confusion among customers but less confusion, although certainly disagreement among scientists. Therefore, due to the uncertainty surrounding the most appropriate tools and applications of MST, a workshop was convened in San Antonio, Texas over a three-day period (February 16-18, 2005).

The overall objective of the workshop was to identify the knowledge gaps and research needs for application of MST technologies by the wastewater and drinking water industries. Some of the specific questions included:

- Is source tracking a useful approach for watershed management?
- Are the most appropriate methods being used in MST projects?
- What are the most appropriate methods?
- Are some of these methods ready for standardization?
- What are the methodological research needs?
- What are the most suitable targets? (chemicals, organisms, genes)
- Where are the information gaps?
- How should future method comparison studies be designed?
- What level of resolution is needed for discriminating between sources of contamination?

In addition to the three member project team, seven member project scientific committee (PSC) and WERF project manager, workshop participants included 36 individuals representing water and wastewater utilities, academia, government agencies (federal and state), medical institutions, and private laboratories from the United States, Canada, United Kingdom, Australia, and New Zealand. Collectively, these individuals had expertise in the microbiology of water and wastewater, microbial ecology, biological data analysis, microbial source tracking, environmental microbiology, bacterial population genetics, and clinical epidemiology. The workshop was also attended by 75 observers representing local (Texas) water agencies, state government agencies, commercial laboratories, and academia. Presentation topics included a report from a 2002 MST workshop, the assumptions and limitations of MST, the confounding role of sediment in MST, application of culture-independent methods, population genetics and ecology of *E. coli*, comparison of methods, applications of library-independent chemical methods, case studies in marine and freshwater systems, and an overall European perspective of MST. Breakout groups discussed issues centered on the selection of the most appropriate organisms and targets for MST, study design criteria and standardization, and information handling and data analysis. The workshop comprised a combination of technical presentations, breakout groups, and open discussions, as detailed below. The role of the workshop organizers was to remain as neutral facilitators of the meeting and synthesize the outcome of the workshop.

Workshop Agenda

Wednesday February 16, 2005

9:00 – 9:10	Welcome	Patrick Hacker, Metropolitan Water District of Southern California (MWDSC)
9:10 – 9:30	Objectives and Logistics	Paul Rochelle, Metropolitan Water District of Southern California
9:30 – 10:00	Identifying the issues and the workshop process	Ricardo De Leon, Metropolitan Water District of Southern California
10:00 – 10:30	Issues and outcomes from a 2002 source tracking workshop	Stephen Weisberg, Southern California Coastal Water Research Project Authority
10:30 – 11:00	Assumptions and limitations of MST methods	Jorge Santo Domingo, U.S. Environmental Protection Agency (U.S. EPA)
11:15 – 11:45	Understanding the role of sediment in BST	Peter Hartel, University of Georgia
11:45 – 12:15	Culture independent methods	Katharine Field, Oregon State University
12:15 – 12:45	Case studies in South Carolina watersheds	Geoffrey Scott, National Oceanic and Atmospheric Administration
13:45 – 17:45	Workgroups:	**Workgroup leaders**
	Selection of targets?	Ricardo De Leon, MWDSC
	Selection of methods-I?	Sandra McLellan, Great Lakes WATER Institute
	Standardization	Jorge Santo Domingo, U.S. EPA
	Study design-I?	Kenneth Hyer, U.S. Geological Survey

Thursday February 17, 2005

8:00 – 9:00	Reports from Wednesday workgroups	Workgroup leaders
9:00 – 9:30	*E. coli* population genetics	David Gordon, The Australian National University
9:30 – 10:00	Methodological issues in MST studies: Relative comparison of phenotypic and genotypic methods	Mansour Samadpour, Institute for Environmental Health
10:00 – 10:30	Library-independent techniques and chemical methods	Nicholas Ashbolt, The University of New South Wales
10:45 – 11:15	Bacterial source tracking in Lake Superior streams- A case study	Randall Hicks, University of Minnesota-Duluth
11:15 – 11:45	A European perspective	Huw Taylor, Brighton University
11:45 – 12:30	Open microphone	
14:30 – 17:30	Workgroups:	**Workgroup leaders**
	Study design-II	Robin Oshiro, U.S. EPA
	Information handling	Tom Edge, Environment Canada
	Immediate and long-term needs	Ricardo De Leon, MWDSC
	Selection of methods-II?	Nicholas Ashbolt, University of New South Wales

Friday February 18, 2005

8:30 – 9:30	Reports from Thursday workgroups	Workgroup leaders
9:30 – 11:00	Information and research gaps	Ricardo De Leon, MWDSC
11:00 – 12:00	The future of microbial source tracking	Paul Rochelle, MWDSC
13:00 – 15:00	All participants panel discussion and open microphone	
15:00 – 15:30	Workshop wrap-up and adjournment	

CHAPTER 3.0

UPDATES BASED ON WORKSHOP PRESENTATIONS, RECENT LITERATURE AND OTHER WORKSHOPS

MST related research is ongoing and new information ensures that reviews rapidly become outdated. This chapter provides updates to Chapter 1.0 with information provided during the workshop, in selected recent publications, and at meetings held since the San Antonio workshop. A recent U.S. EPA document provides guidance to end-users considering MST technologies for assessing water quality and TMDL-related activities (U.S. EPA, 2005). It provides a summary of methods, discusses assumptions and limitations of the methods, and provides examples of real-world applications of many techniques, including ARA, rep-PCR, PFGE, ribotyping, coliphage genotyping, and host-specific Bacteroidetes markers. Consulting this guide should be part of the study design process when planning to conduct MST studies.

Topics covered by plenary presentations at the San Antonio workshop are detailed in Table 3-1. Prior to 2002, the MST field was largely comprised of independent researchers and method developers using a variety of techniques and working mostly in isolation. As described by Stephen Weisberg, a consequence of a 2002 MST workshop was enhanced communication between many researchers and implementation of collaborative studies. The multi-investigator method evaluation studies that were conducted as a consequence of this earlier workshop all reached similar conclusions (Griffith et al., 2003; Stewart et al., 2003; Stoeckel et al., 2004). Most, but not all, methods achieved better than random chance identification of host animals. Genotypic methods generally outperformed phenotypic methods but the biggest problem for most methods was the high rate of false positives and the high rate of failure to classify isolates to any source. Also, different outcomes were generated from the same data set, depending on the particular pattern-matching algorithm that was used. Host-specific PCR showed the most promise in the collaborative evaluations but was not quantitative and was only developed for two sources at the time. An additional methods comparison study that evaluated ribotyping and ARA of *E. coli* and enterococci with blinded proficiency samples concluded that the methods are not sufficiently mature or accurate for application to environmental samples (Moore et al., 2005). Discussion at the San Antonio workshop centered on the reality check that these method comparisons provided for the MST community. Despite the many MST workshops held in recent years, there was still no clear message to potential clients, going into the San Antonio workshop. Methods comparison studies have caused significant concerns among some potential end-users over the wisdom of performing applied studies. Therefore, considering the current state of the science, MST end-users should be informed that the tools are still research methods that may fail to generate results that can be readily interpreted. The workshop provided an opportunity to define the steps necessary for further optimization, method development, and evaluation.

Table 3-1. Technical Presentations During Plenary Sessions.

Title	Presenter
Issues and outcomes from a 2002 source tracking workshop	Stephen Weisberg Southern California Coastal Water Research Authority
Assumptions and limitations of MST methods	Jorge Santo Domingo U.S. Environmental Protection Agency
Understanding the role of sediment in BST	Peter Hartel University of Georgia
Culture-independent molecular methods for fecal source identification	Katharine Field Oregon State University
Case studies of microbial source tracking in South Carolina watersheds	Geoffrey Scott National Oceanic and Atmospheric Administration
Population genetics and ecology of *Escherichia coli*	David Gordon Australian National University
Microbial source tracking: Identification and tracking of the sources of microbial pollution	Mansour Samadpour Institute for Environmental Health
Library-independent chemical and microbial methods: An example tool-kit	Nicholas Ashbolt University of New South Wales
Bacterial source tracking in Lake Superior streams- A case study	Randall Hicks University of Minnesota-Duluth
A European perspective	Huw Taylor Brighton University

The scale of the fecal contamination problem was highlighted by Jorge Santo Domingo who provided an estimate of 1×10^{12} kg of human, pet, and livestock feces produced each year in the United States, 96% of which is produced by just three groups of animals (cattle, pigs, and chickens). MST is predicated upon a number of assumptions, the primary one being that the intestinal microbial flora of various animal groups are different and this difference can be discerned with an appropriate tool. The rationale for the assumption of unique intestinal flora is that temperature, diet, and physiology of the digestive system, along with competition for space and nutrients leads to selection of host-specific flora. Also, in recognizing the need for the development of new tools, Jorge also described a hypothetical microarray that might be developed to simultaneously detect pathogens and indicator microorganisms and track the source of contamination. Although the application of microarrays to environmental samples is in its infancy, proof of concept has been demonstrated for wastewater analysis (Berthiaume et al., 2005).

Katharine Field presented a synopsis of culture independent host-specific PCR methods concentrating on detection of 16S rDNA markers in Bacteroidales (*Bacteroides-Prevotella* group), the most rigorously tested of this type of method. Bacteroidales account for a significant portion of the fecal flora of human and animal feces (more so than *E. coli*). They are anaerobic, occurring naturally only in body cavities and have limited survival upon release to the environment. Cultivated strains do not adequately represent fecal diversity and the most common fecal Bacteroidales remain uncultivated. Recent advances have led to the development of human, bovine, porcine, equine, elk, and dog strain specific PCR primers. Moreover, cow and human markers demonstrated geographic stability (Dick et al., 2005a). Almost all Bacteroidales in ruminant feces represent new uncultured strains (possibly novel genera), which led to the development of ruminant-specific primers (Dick et al., 2005a,b). The technique has also been adapted to a high-throughput 96-well format. Methods using surrogate organism spikes to assess the efficiency of DNA extraction and the level of PCR inhibition are also being developed. There has also been progress towards quantification with the development of a real-time quantitative PCR (Q-PCR) assay targeting a human-specific *Bacteroides* 16S rDNA marker (HF183; Seurinck et al., 2005), although the *Bacteroides* species/strain that carries this marker has not been cultured. In a field study using Bacteroidales-specific PCR, water samples that were classified as containing human fecal pollution were from sites associated with sewage treatment plants whereas while sites with ruminant fecal pollution were associated with concentrated animal feeding operations. While promising, it is necessary to evaluate the technique with a broad range of human samples since there is considerable host-dependent diversity in the composition of human *Bacteroides* communities (Pang et al., 2005).

Other methods based on PCR amplification of host specific markers include detection of toxin encoding genes in enterotoxigenic *E. coli* (STII), virulence genes in enterococci (*esp*), and 16S rDNA markers in bifidobacteria (Khatib et al., 2002; Scott et al., 2005; Bonjoch et al., 2004). QPCR assays for additional virulence genes in enterococci (Vankerckhoven et al., 2004) and potentially human-specific *Clostridium* spp. that account for ~5% of human fecal microflora (Rousselon et al., 2004) may also be useful. Issues that still need to be addressed for many of these markers include assessing geographic and temporal stability, relatively rare occurrence of pathogens, low sensitivity requiring enrichment culture of some organisms, and lack of markers for diverse animals (rather than just humans). However, there was general agreement in San Antonio that these methods have a lot of promise and increased effort should be focused on discovering a broader range of host-specific markers.

Geoff Scott described the use of ARA, ribotyping, and coliphage genotyping to investigate potential sources of contamination in South Carolina watersheds, estuaries, and coastal waters. Of 235 surface water *E. coli* isolates, 41% matched a library isolate with a 90% similarity threshold, and of these, cattle and raccoons accounted for 96% of the isolates. Overall, the results indicated that non-human inputs were the primary sources of contamination (Table 3-2) but the of origin type I coliphages in municipal sewage is unclear.

Table 3-2. Relative Isolation of Coliphage Genotypes from Environmental and Animal Samples*.

		Coliphage genotype				
	N	I	II	III	IV	Not typed
All surface waters	534	93.8	4.3	1.1	0.0	0.7
Catawba R.	219	92.7	5.0	1.8	0.0	0.5
Saluda R.	193	95.3	2.6	1.0	0.0	1.0
Savannah R.	121	93.4	5.8	0.0	0.0	0.8
Municipal wastewater	133	9.8	14.3	71.4	0.0	4.5
Hog Lagoons	55	67.3	0.0	0.0	5.5	27.3
Chickens	94	51.1	0.0	0.0	48.9	0.0

*(Data courtesy of Jill Stewart and Geoff Scott, NOAA; *submitted for publication*).

Randall Hicks also described case studies in which ARA and rep-PCR with the BOXA1R primer were used in MST investigations in Lake Superior watersheds. These studies included land-use surveys to determine whether there was a correlation between sources of *E. coli* (as determined by MST) and land use at varying spatial scales. With a 90% similarity cut-off, BOX-PCR assigned 51% of surface water isolates (N = 590) to a source category compared to only 1% for ARA. Although human, pet, and farm animal inputs were identified, the primary sources of surface water contamination appeared to be wildlife and waterfowl. A weak positive relationship was identified between the relative abundance of waterfowl *E. coli* and the percentages of agricultural and grassland areas within the watershed.

Peter Hartel highlighted the importance of sediments and discussed their role when interpreting MST data. During baseflow conditions, large numbers of fecal bacteria are found in sediment and *Ent. faecalis* isolates with different ribotype patterns survive differently. During windy conditions, sediments in shallow water are disturbed and bacterial counts increase, even in the absence of rain. Anything that disturbs sediment may increase bacterial counts. Also, during runoff conditions, sediments and outside fecal sources are disturbed leading to increased counts. When water levels drop, sediments become desiccated; when rewetted, fecal enterococci, but not *E. coli*, appear to survive and regrow in some sediments. Due to survival, persistence, and growth of bacteria in sediments, they provide an environmental reservoir. Therefore, depending on the conditions, sediments have a bacterial "memory" that can affect MST results and interpretation. This supported by the high levels of *Enterococcus* spp. in inter-tidal sediments indicating retention and regrowth (Ferguson et al., 2005). According to some estimates, there may be up to 65% overlap between bacterial populations in sediment and the overlying water column at any given time.

Due to its variable and heterogeneous population structure, it has previously been suggested that *E. coli* is a poor target for MST (Gordon, 2001). David Gordon expounded further on this issue by discussing the ecology and population genetics of *E. coli*. The prevalence of *E. coli* in birds is associated with increasing size of the birds or with a commensal association with humans. Strains of *E. coli* can be split into four general groups that differ markedly in their genotypic and phenotypic characteristics. Groups A and B1 occur in all vertebrates, group B2 predominates in animals that have microbial fermentation processes in their hindgut, and group D strains occur in warm-blooded vertebrates (Figure 3-1). Groups A and B1 may be free-living strains in water.

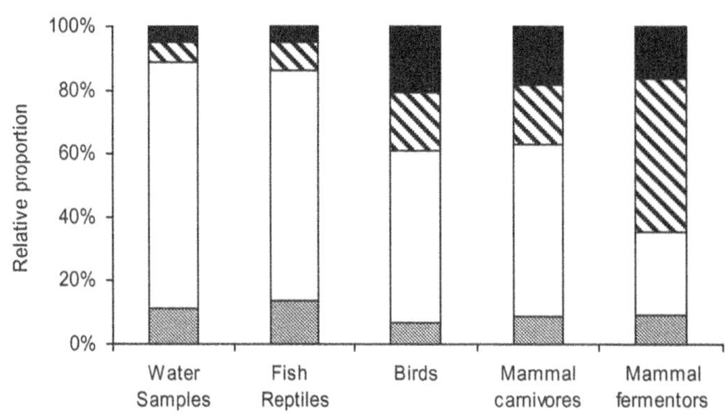

Figure 3-1. Distribution of *Escherichia coli* in Various Types of Sample. Strains can be classified into groups A (grey), B1 (white), B2 (striped), and D (black). Data courtesy of David Gordon (Gordon and Cowling, 2003; and unpublished).

Temporally stable strains of *E. coli* that are responsible for seasonal blooms containing >100,000 cfu/L in widely dispersed locations across Australia cannot be matched with database fecal isolates according to MLST or capsule gene analyses (Power et al., 2005). Bloom dynamics are not correlated with fecal contamination and the health risk of these particular strains of *E. coli* is currently unknown. This example illustrates the importance of potentially free-living *E. coli* strains that are capable of environmental proliferation and suggests a possible confounding effect on *E. coli*-based MST methods. Differential survival of *E. coli* phylotypes also casts doubt on the utility of this organism as a target for MST (Anderson et al., 2005).

Based on *Eco*RI and *PvuI*I ribotyping of 262 *E. coli* stream isolates, only 3.6% of isolates were common during base-flow and storm-flow conditions (Hartel et al., 2004). In addition, the majority of ribotype patterns were not shared between different stream locations. Although landuse patterns and flow conditions may have affected the distribution of *E. coli* isolates, the results may have been skewed by the relatively low number of isolates that were analyzed. The importance of a sufficiently large source library was emphasized by Mansour Samadpour who discussed the epidemiological approach to MST based on ribotyping and 100% matching of profiles for source identification.

Huw Taylor described a multi-investigator, multi-national evaluation of source tracking methods with the aim of establishing one or a few easy procedures applicable to different geographic regions in Europe. Methods that proved to be most useful were the presence of type II coliphages, coprostanol, and coprostanol:coprostanol+epicoprostanol ratio (Blanch et al., 2004). Other tools currently undergoing evaluation include sorbitol fermenting bifidobacteria and detection of bacteriophages infecting *Bacteroides thetaiotaomicron.*

Prior to the San Antonio workshop, there appeared to be an overall understanding that chemical methods were somewhat limited for source tracking. In contrast, Nicholas Ashbolt presented a strong case supporting library-independent chemical methods, demonstrating that fecal sterol fingerprinting is relatively advanced. By the end of the workshop there was a consensus in favor of continued development and evaluation of chemical methods, particularly fecal sterols. Of the C_{27} stanols, 5β-cholestan-3β-ol (coprostanol) is a human marker and high levels indicate relatively fresh feces. The C_{29} stanols, 24-ethylcoprostanol and 24-ethyl-*epi*-coprostanol are herbivore fecal biomarker. Chromatography of fecal samples typically yields 24 stanols per trace with animals differentiated by relative peak differences (Figure 3-2). If the ratio of coprostanol to 24-ethylcoprostanol is >3, fecal contamination is presumed to originate solely from humans. A ratio <0.4 indicates a solely herbivorous contribution. While interpretation of relative contributions can be hindered by background levels of various sterols and stanols, development of multi-stanol profiles (Faecalprints) increased the utility of the method. A field study utilizing the Faecalprint technique concluded that intensive agricultural development in an area accounting for <5% of the watershed lead to a doubling in overall pathogen risk. However, sterol profiling is not proposed as a stand-alone technique; multiple methods are necessary for MST assessments. The complementary use of a suite of microbial indicators and fecal sterols for managing non-point source pollution has been termed “environmental forensics”.

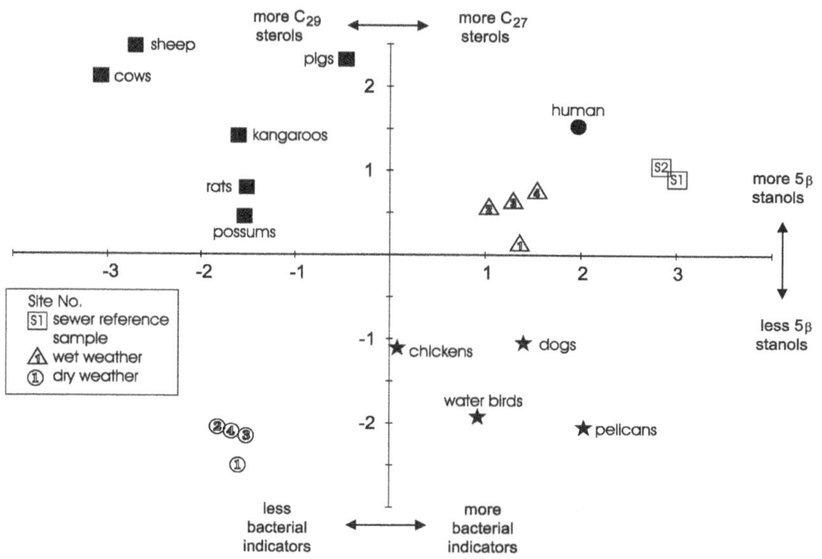

Figure 3-2. Principal Component Analysis of Human and Animal Fecal Sterols. (Adapted from Leeming et al., 1998).

Cocaine and its primary urinary metabolite were detected by mass spectrometry in municipal wastewater treatment plants and surface waters in populated areas (Zuccato et al., 2005) suggesting that illicit drugs may be additional indicators of contamination related to human activity. Non-specific chemical methods may also have a role in future source tracking efforts. Treated wastewater effluents have been analyzed by UV absorbance and fluorescence, ^{13}C and ^{1}H nuclear magnetic resonance spectroscopy, gas chromatography, and pyrolysis-gas chromatography/mass spectrometry (GC/MS). These studies have generally demonstrated that effluent derived organic matter (EfOM) has a different composition than natural organic matter (Sirivedhin and Gray, 2005). Therefore, these spectral techniques have the potential to identify the presence of municipal sewage effluent in water. Combining conventional water quality parameters (pH, dissolved and total organic carbon, turbidity, specific UV absorbance) with pyrolysis-GC/MS data analyzed by principal component analysis, demonstrated that the organic quantity and quality of treated wastewater effluents were distinct compared to pristine waters (Sirivedhin and Gray, 2005). Benzaldehyde, benzonitrile, chlorobutanoic acid, and furancarboxaldehyde were indicators of anthropogenic or wastewater sources of contamination. EfOM discharged by a wastewater treatment plant was detected and a long way downstream of the plant and had a large impact on downstream water quality. Further refinement of these techniques may eventually allow categorization of fecal contamination from additional sources.

3.1 Toronto MST Workshop

As a follow-up to the San Antonio workshop, and as an example of cross-workshop coordination and cooperation, Dr. Rochelle attended a workshop in Toronto, Canada three weeks after the San Antonio meeting. This workshop was entitled "Microbial Source Tracking in Aquatic Ecosystems: The State of the Science and an Assessment of Needs" and was organized by Tom Edge and Karl Schaefer (National Water Research Institute, Environment Canada). San Antonio participants that also attended the Toronto workshop were Tom Edge, Michael Jenkins,

Cindy Nakatsu, Valerie Harwood, and Nicholas Ashbolt. Many of the issues discussed were the same as those addressed in San Antonio. However, additional valuable information was presented that was not covered at the San Antonio workshop. A rural Canadian perspective presented at the Toronto meeting was that MST tools are needed on a very small/local scale where many individual farmers or homes have chronically contaminated single property wells. It is hoped that MST can be used to determine the source of contamination in these wells.

The use of specific pathogens for MST, particularly *Cryptosporidium* spp., was discussed in Chapter 1.0. This approach has been greatly expanded by Norman Neumann (Alberta Provincial Laboratory for Public Health). In two independent studies using *Cryptosporidium* genotyping, the many of the oocysts were of wildlife origin with a significant contribution by cattle (*C. andersoni*) during a period of heavy precipitation. In these instances, targeting a specific pathogen proved useful because both rivers were heavily impacted by agriculture and other activities, indicating a high likelihood of detecting *Cryptosporidium* oocysts. However, as stated in Chapter 1.0, in many watersheds particular pathogens may not be present in most samples, even in the presence of fecal contamination.

MST microarrays targeting the 16S rRNA and *cpn*60 genes have demonstrated discrimination between some human and animal samples but considerably more work is necessary to develop microarrays into a reliable MST tool (Roland Brousseau, National Research Council of Canada). Current microarray technology limits their sensitivity for use with environmental samples without prior target amplification. Nevertheless, the processing power of DNA microarrays, with the ability to simultaneously interrogate target sequences with hundreds or thousands of probes, makes them an attractive tool warranting further investigation (Berthiaume et al., 2004).

Pierre Payment (Institut National de la Recherches Scientifique) reported the development of a novel approach to source tracking by moving away from microorganisms. This method involves the detection of animal-specific mitochondrial DNA sequences in surface water and is based upon the finding that animal feces contain large amounts of cellular debris from blood and intestinal cells. PCR primers specific for human, bovine, ovine, and porcine mitochondrial DNA were used singly and in multiplex reactions and detected at least one of the four target species in untreated domestic sewage, agricultural run-off, pig farm effluents, and river water (Martellini et al., 2005). Since each target species generated amplicons of unique size, gel electrophoresis suffices as the assay end-point, although hybridization with an internal probe or amplicon sequencing would be a necessary confirmatory step.

The application of multiple methods to identify sources of fecal pollution in water was discussed by Charles Hagedorn (Virginia Tech). Optical brighteners that are typically included in laundry detergents can be detected by fluorimetry with excitation at 360 nm and fluorescence at 410-430 nm. Detection of such compounds can be used as a fast presence/absence screen for sewage or other recent anthropogenic sources of contamination, but the method requires additional standardization and calibration before it can be routinely adopted. In some instances, ARA indicated the presence of a human signature, which was confirmed by PFGE of *E. coli* isolates. Continuous, boat-mounted fluorimetry of optical brighteners was then used to identify a plume of contamination and trace the contamination to its source by sailing up the concentration gradient. Some of these sources were individual riverside homes with leaking septic tanks or apartment complexes with illegal sewer connections to storm drains. These case studies clearly indicated the potential of MST when a combination of methods was used in relatively small, defined locations. The optical brightener results also add to the conclusion from San Antonio that it would be worthwhile revisiting some of the chemical methods.

Some of the discussions in San Antonio focused on the acceptable and achievable level of resolution in apportioning sources of contamination. While some participants felt that human versus non-human discrimination is the most useful, others insisted that discrimination to the individual animal species is necessary. In Toronto, Tom Edge explained that wildlife, particularly birds, can have a significant impact on water quality, so wildlife needs to be a consideration in apportioning fecal pollution sources. Therefore, a three-way discrimination between humans, agricultural/domestic animals, and wildlife may be the minimum acceptable level of resolution. However, particular challenges of applying MST to wildlife populations include temporal/seasonal loading complications, acquisition of fecal samples, and assessing the human health risk from microbes in wildlife. Also, mitigation efforts can be challenging if wildlife is determined to be a significant contributor of contamination; wildlife populations can be difficult to control and there is a need to balance riparian habitat needs against facilitating wildlife fecal loadings into streams.

In summarizing the Toronto workshop, Tom Edge made the following points, all of which are in agreement with conclusions from the San Antonio workshop:

- There is no silver bullet method.
- There are many methods but none of them are standardized. Some methods are much further along in their development compared to others but the field continues to evolve.
- More work is needed on assessing and measuring the robustness, reliability, accuracy, resolution, and sensitivity of the methods.
- There is no clear indication as to the most appropriate tools.
- It will be necessary to combine methods.
- If end-users cannot wait for the science to mature, it will be necessary to incorporate extensive QA/QC into MST studies.
- There are examples of case studies that demonstrate the utility of some MST methods.
- It is important to manage end-user expectations so that realistic and attainable objectives are set.

CHAPTER 4.0

REPORTS FROM BREAKOUT GROUPS

Each of the workshop participants were involved in two different breakout groups over the course of the workshop (Figure 4-1). A volunteer group leader facilitated the discussions of each breakout group with a note-taker selected from within each group. The workshop organizers provided a list of questions and discussion topics for each breakout group but the groups were also able to discuss additional issues and formulate their own questions. While some groups addressed the specific questions presented at the start of each session, others had more general discussions. The following sections summarize the deliberations of each of the breakout groups. Although each breakout group discussed individual topics, it quickly became apparent that all of the topics were inextricably linked and it was therefore difficult to address many of the issues in isolation. Inevitably, each group had questions and raised issues that were also considered by other groups.

The broad topics covered by the eight breakout groups were:

1. Selection of Methods (two groups)
2. Selection of Targets
3. Appropriate Study Design (two groups
4. Standardization
5. Information Handling
6. Immediate and Long-Term Research Needs

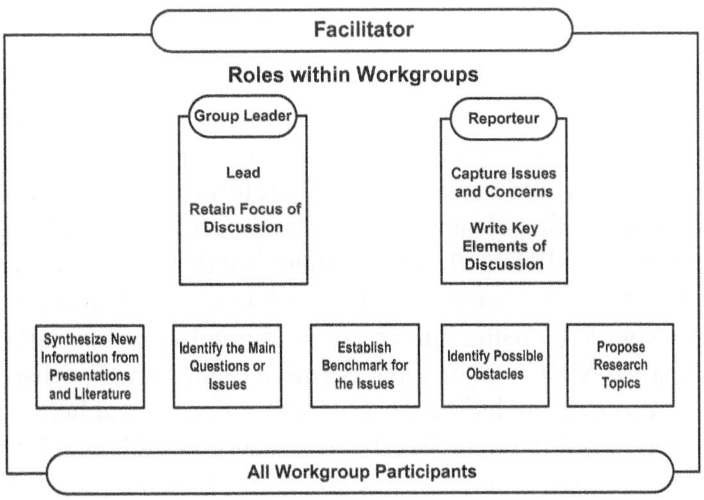

Figure 4-1. Roles and Responsibilities in Workgroup Sessions.

4.1 Selection of Methods

Table 4-1. Methods Breakout Group 1 Members.

Sandra McLellan (group lead)	
George Di Giovanni	Mark Borchardt
Peter Hartel	Rudolph Ellender
William Hamilton	Rick Meinersmann
Mansour Samadpour	James Johnson

Questions Presented to the Breakout Group

- Which methods are currently in use and what are their respective advantages and disadvantages?
- Which are new or emerging methods for source tracking and what are their respective advantages and disadvantages?
- Could method combinations be used to mitigate disadvantages of a single method?
- What method combinations which have the potential to enhance detection capability?
- Should technology used in other fields be considered, and if so, which technology?

The first methods subgroup also asked the question: How does source information translate into human health risk? Knowing the answer to this question would provide a better framework for assessing methods. However, there is currently little information on how the results from MST methods translate into health risk.

Selecting a method or combination of methods is difficult when presented with a list of all available methods. No single method or suite of methods is likely to be the most appropriate for all possible scenarios. The choice of method and the relative strengths and weaknesses methods are very dependent on the particular question that is being asked. For example, the ideal method for differentiating between human and non-human contamination at a beach will probably not be the most suitable for apportioning the relative contribution of animals and humans in dealing with watershed management. Therefore, it is not possible to select the most appropriate methods without first clearly defining the information needs and objectives of the end-user. The application of MST methods will therefore depend on end-users needs and resources. The first level of application would be to determine whether contamination has a human or non-human source by using human and/or animal-specific markers. This could be followed by more detailed MST methods that permit a higher level of discrimination, depending on the availability of resources including the proximity of technical help, level of funding, time-frame, and the intended use of the data. Finally, end-users should complete an analysis of the total picture and select the most suitable remedial action. In lieu of selecting methods, the group discussed the importance of adopting a practical approach to field applications of MST. This should be a multistage process that:

- Collects existing data concerning the study site (e.g., topographical maps, historical data of indicator contamination, sewage spills)
- Identifies key personnel (geologist, microbiologist, hydrologist, ecologist, etc)
- Identifies hot spots (high and frequent contamination by selecting multiple sampling locations in problematic locations and frequent sampling)
- Determines the level of resolution that is needed to apportion sources:
  -Human versus non-human.
  -Human, agricultural animals, wildlife.
  -Human, rat, dog, cat, cattle, and the proportions of other animals.

For all methods, more research is necessary to improve accuracy, precision, specificity, and robustness. Specific recommendations included:

1. Assess geographical specificity of current host-specific gene targets (those that discriminate between humans and non-humans and those specific for particular animals).
2. Evaluate inter-laboratory reproducibility of current host-specific gene targets.
3. Develop QA/QC and standardization procedures for current markers.
4. Validate current gene targets and identify additional host-specific markers that may be useful for MST.
5. Develop quantification methods for host-specific molecular markers and determine how well these methods compare to the resolution of library-based methods.
6. Library based methods should not be discarded. These methods received the most support from group members when information is needed beyond human versus non-human discrimination. However, a thorough comparison study of library-based methods needs to be performed.
7. Investigate environmental persistence of all selected targets/markers.
8. Conduct further investigations of chemical methods to determine the biodegradability, persistence, and relation to benchmark indicators of chemical targets.
9. Perform multi-variant analysis of method combinations. While individual methods may not stand alone some combinations of methods might prove useful but no thorough multiple method comparisons have been performed.
10. Conduct prospective studies on the effectiveness of remediation strategies.
11. Perform epidemiological studies to assess the health risk in relation to particular sources of contamination.

Table 4-2. Methods Breakout Group 2 Members.

Nicholas Ashbolt (group lead)	
Geoffrey Scott	Menu Leddy
Jan Vinje	Randall Hicks
Jennifer MacDonald	Marylynn Soule
Valerie Harwood	Bruce Wiggins

The second methods breakout group discussed the importance of providing methods to support TMDL requirements with the primary focus being on the impact from wastewater discharges and water reuse. From this perspective, it is important that MST methods provide a link to current microbial indicators. The group tabulated 13 criteria for the most commonly used MST techniques. This exercise was completed for the majority of methods listed in Table 1-1 in an attempt to gage the level of confidence in each of the techniques. Representative examples are provided in Table 4-3. Some methods such as t-RFLP, RAPD, and vibrational spectroscopy were considered too immature or inappropriate for consideration at this time. New/novel methods that may be worth considering include detection of detergents and surfactants, hormones, and bile salts. Pyrolysis-based spectral analyses and matrix assisted laser desorption/ionization time of flight (MALDI-TOF) mass spectrometry of compounds in water may also be useful.

The group agreed with most of the recommendations of group 1. There is no silver bullet so method combinations are considered the way forward, but the particular combination will depend on the questions being addressed. The toolbox should contain at least one microbial method and one chemical method that have been validated. Methods should be complementary but unrelated to increase the weight of evidence. The toolbox of methods should be evaluated in multiple laboratories with samples covering urban and agricultural watersheds. However, the second group considered library-based methods too time consuming and too expensive to be practical. In addition, the instability of libraries over time may limit their utility. A U.S. EPA 2003 workshop also concluded that library-based approaches do not represent a long-term solution to microbial source tracking requirements.

Table 4-3. Assessment of Relative Performance of Some MST Methods.

Criteria	Culturable viruses	Non-culturable viruses	Optical Brighteners	Sterols	rep-PCR *E. coli*	rep-PCR Enterococci	Ribotyping *E. coli*	Ribotyping Enterococci
Reproducibility between replicates	✓+	✓+	✓	✓	✓	✓	✓+	✓+
Specificity (false + or -)	✓+	✓+	✓?	✓	✓-	✓	✓- (1 enzyme) ✓ (2 enzymes)	✓ (2 enzymes)
Detection limit	?	?	-	✓	✓+	✓+	✓+	✓+
Accuracy for source	✓+	✓+	?	✓?	✓-	✓	✓	✓
Direct association with health risk	✓+	✓+	-	-	-	-	-	-
Geographic applicability	✓+	✓+	✓	✓	-	-	-	-
Temporal stability	✓	?	✓?	✓+	-	?	-	?
Maturity of method	✓	✓	✓?	✓	✓+	✓+	✓+	✓+
Potential for quantification	✓	✓	✓	✓	✓	✓	✓	✓
Level of host discrimination	Human,?	Human, ?	Human	Human, Herbivores, Birds	Human, Pet, Livestock?	Human, Pet, Livestock?	Human, Pet, Livestock?	Human, Pet, Livestock?
Occurrence of target	Sewage✓, Rare in feces	Sewage✓, Rare in feces	✓?	✓+	✓+	✓+	✓+	✓+
Relationship to current indicators	-	-	✓ (sewage)	✓	✓+	✓+	✓+	✓+
Cost, logistics, rapidity, level of expertise, required resources	Concentration method, matrix effects?, requires cell culture	Concentration method, matrix effects?, requires PCR facility	Fluorometer HPLC	$300/ sample, GC/MS	Cheap	Cheap	Expensive	Expensive

✓+ Very good
✓ Good
✓? Good but more evaluation necessary
✓- Could be better
? Unknown
- Weak

4.2 Selection of Targets

Table 4-4. Targets Breakout Group Members.

Ricardo De Leon (group lead)	
Huw Taylor	Valerie Harwood
Jill Stewart	David Gordon
Katharine Field	Marylynn Soule
Robin Oshiro	Jennifer MacDonald

Questions Presented to the Breakout Group

- Which targets, microbial and chemical, could be considered sufficiently meritorious and established for source tracking studies?
- Are all of these targets equal? How can they be ranked or further classified?
- Which emerging targets are worth considering and why?
- Identify target gaps in MST.
- What approach(es) or criteria could be developed to select new targets for MST?

The group discussed some of the additional factors, other than simply method performance, that influence the selection of source tracking target. For example, different regulatory or health impacts will require selection of different targets. Also, the particular application or end-use (e.g., TMDL, recreational use, etc) will probably require different source tracking targets. In addition, target selection criteria will be different for library dependent or library independent methods.

Emerging targets or processes that require further investigation include:

- Metagenomics as a process for target selection.
- Source-specific pharmaceuticals for human and domesticated animals.
- Source-specific biomarkers.
- Virulence factors in human-specific enterococci.

Benchmarks that need to be considered when selecting a method are:

- Tier 1 targets/methods should discriminate between human and non-human sources.
- Tier 2 targets/methods will differentiate between individual animal sources.
- Define the geographical area or domain that is applicable to particular targets.
- Define the characteristics or attributes of the scenario under investigation as part of target selection process.
- Host specificity, survival and transport in the environment needs to relate to the issue under investigation.
- The target needs to be focused on MST, as opposed to readily measured public health impacts.

Possible obstacles to the successful implementation of MST methods include adopting an overly ambitious areas/region of study. Requests for proposals and their responding proposals should initially limit the geographical area of study. This is particularly critical for library-dependent methods. Also, targets may be limited by certain geographical areas and by host populations and their social behavior.

Recommended criteria for evaluating the utility of targets are:

- Host specificity
- Level of discrimination
- Persistence in the environment—just the right amount—not too long/short
- Broad distribution in hosts
- Geographic stability
- Temporal stability
- Marker stability/reproducibility
- Differential survival

Recommended research topics include:

1. Validate biomarkers (e.g., virulence factors) for source contamination using the target evaluation criteria above to help characterize the ideal target.
   -Applicable to various situations (i.e., TMDL, beaches, shellfish growing waters, sourcewater protection) where MST would be useful.
2. Apply tools such as metagenomics to develop host specific biomarkers as potential source specific targets.
3. Develop practical (e.g., not pie in the sky) approaches to simultaneously query multiple targets.
4. Look at the predictive relationship or co-occurrence between markers and pathogens.
5. Develop a ranking process for selecting targets under various applications.
6. Determine the association between human health outcomes and specific sources of fecal contamination.

4.3 Appropriate Study Design

Table 4-5. Study Design Breakout Group 1 Members.

Kenneth Hyer (group lead)	
Stephen Weisberg	Nicholas Ashbolt
Samuel Myoda	Christopher Kitts
Jan Vinje	Jayson Wilbur
James Kern	Bruce Wiggins

Questions Presented to the Breakout Group

1. From a study design perspective, what are the different types or categories of study design?
2. What are possible key variables for each type or category of study design?
3. What are key guidelines for each type of study design?
4. What are the impacts of method on sampling scheme?
5. What are possible confounding factors in source tracking studies and suggested control measures?

This group discussed the relevance of study design and concluded that additional method evaluation studies should be prioritized above research that is aimed at study design issues. These studies should include evaluating different targets, laboratory methods, and data analysis techniques. The methodological variability/error is likely to be much greater than the error induced because of a well though out sampling design. In a given study it is probably better to balance limited dollars by investing in more supporting methods than perfecting the sampling plan.

Nevertheless, there are challenges associated with study design. Certainly, one-size fits all study designs will not work. The design of any field study will depend on the specific study question or scenario under investigation. A tailored, prescriptive study can only be designed once the question is resolved. Samples will need to be collected in such a way that environmentally relevant and interpretable data are generated. Various aspects of the environmental setting/system complexity are critical to the sampling plan: hydrology, land use, climate, soils and geology. Study design is inherently influenced by the target organism(s) selected, method of indicator characterization, and methods of data analysis. Even with the best study design, MST results may be semi-quantitative, at best. Discrimination resolution is also a critical factor in study design:

- Fecal or non-fecal (is it a health risk?).
- Human or non-human.
- Human, domestic, livestock, wildlife (broad categories).
- Specific to individual host species.

Benchmarks for improving study design include:

- Quality assurance and quality control are critical. Based on recent research indicating problems with method, we must do a better job of incorporating QA/QC into studies:
    - -Direct evaluation studies
    - -Studies should use multiple MST methods
    - -Other chemical tracers
    - -Positive controls (spiked samples)
    - -Negative controls (blanks)
    - -Measure of method variability (sufficient replicates)
    - -Many ways of independent data verification
- The burden of proof is on those performing the study.
- Presentation of QA/QC data is important. Quantitative interpretation of these data is even better.
- Proposal funding decisions should consider the principal investigator's ability to demonstrate a strong understanding of these concepts.

Recommended research topics include:

1. Investigate the role of sediments in the fate of sources, indicator organisms, and pathogens. This will help to develop an understanding of the importance of re-growth, re-suspension, and the associated health risks.
2. Assess differential fate and transport of indicator organisms and pathogens after leaving the host, including matrix issues, partitioning, riparian filtering, etc.
3. Evaluate method combinations to achieve a suite of different analyses to provide a weight of evidence approach.
    - -Which are tools of the trade and which are specialty tools?
    - -Organize/prioritize the toolbox.
4. On a mid- to long-range basis, there needs to be a greater understanding of how the data and study results relate to the public health risks.

Table 4-6. Study Design Breakout Group 2 Members.

Robin Oshiro (group lead)	
Sandra McLellan	Douglas Moore
David Gordon	Mansour Samadpour
Huw Taylor	James Johnson
Elizabeth Casazera	Rudolph Ellender
Jill Stewart	

From a study design perspective, there are a variety of study categories that need to be considered:

- Development of existing and new methods.
- Method comparisons.
- Application, e.g.,
  - -Source water protection
  - -Beach closures
  - -Shellfish growing waters
  - -TMDLs
  - -Wastewater point source discharge
  - -Farming/grazing impacts

For library dependent methods it will be necessary to evaluate methods and develop a site specific library using a multi-stage approach:

- Stage 1: Simple known samples
  - -Test simple known samples, e.g., strains isolated from known fecal samples from humans, dogs, birds, etc.
  - -Repeat of the simple known samples above but with sample identities hidden from the researchers. Methods must achieve 90% accuracy to be considered successful.
- Stage 2: More complicated samples
  - -More (different from previous set) of human, dog, bird strains
  - -Different animal fecal samples, e.g., raccoons, elk
- Stage 3: Pseudo field samples
  - -Mixed samples in various percentages that represent the sources
  - -Water samples spiked with fecal mixtures
  - -Sewage (special category scored differently – laboratory may have to build a library from sewage)
- Stage 4: Field test
  - -With a test that has been validated, begin field testing
  - -Sample in a watershed above and below an input source
  - -Samples with more complex inputs
  - -Perform an intervention or mitigation and assess whether the MST method reflects any before/after changed.

Library independent methods should also be evaluated using a multi-stage approach:

- Stage 1: Assess presence/absence of target in multiple sewage samples.
- Stage 2: Detection of target in fecal samples.
- Stage 3: Site study using: -Sites that are known to have the target
-Sites known to NOT have the target

It is important that researchers develop an understanding of the variability of sources in watersheds. In addition, key variables that need to be considered in the design of studies are: appropriate library size, relative die off, geographic size of the study, temporal stability of the target, and quantification.

Possible confounding factors in MST studies include:

- The environmental matrix.
- Pathogen and indicator behavior in the environment.
- Environmental sources or organisms, e.g., sediments, biofilms.
- Field factors not experienced in the laboratory.
- Sources in the field not represented in library.
- Independent free living populations not necessarily from a known source.
- Inhibitors that give false positive/negative results.

Suggested control measures to help minimize the impacts of confounding factors include:

- Develop a QA/QC program that includes known samples.
- Build a library with clones from known sources and include standard controls.
- Develop standard test conditions.
- Perform inter-laboratory comparisons.

Recommended research topics:

1. Examine the value of using multiple methods followed by multi-variant analysis to determine which combination of methods works best.
2. Parallel comparative and combinatorial assessment of multiple source tracking methods in an analysis of a real-world field problem (e.g., beach closure, TMDL, etc.). The determination of which single method or combination of methods works best would lead to a useful answer.
3. Compare library independent MST methods for their ability to identify a dominant pollution source in blind field tests
4. A three part project should:
    a. Decide criteria for acceptable performance of a method.
    b. Determine how to test for acceptable performance (e.g., how many blind samples are needed, etc.)
    c. Independent testing of submitted methods for accuracy, based on criteria developed in a) and b) above.
5. Develop a better understanding of the behavior of selected targets in the natural environment, e.g., investigate the influence of regrowth, survival in biofilms and sediments, predation, re-release after storm events, and community structure changes in the environment.

4.4 Standardization

Table 4-7. Standardization Breakout Group Members.

Jorge Santo Domingo (group lead)	
Menu Leddy	Cindy Nakatsu
Jeffrey Ram	Randall Hicks
Tom Edge	Marion Savill
Geoffrey Scott	Douglas Moore

Questions Presented to the Breakout Group

1. What is involved in standardization of source tracking methods?
   - Specific field sampling protocols
   - Reference standards and controls
   - Reference sample preparation
   - Standardization of performance standards based on a sound statistical approach
   - Need to establish neutral sites for testing methods performance and accuracy (e.g., U.S. EPA-Environmental Technology Verification Program)

2. Which aspects differ from other standardization procedures for routine monitoring methods?
   - Needs depend on the method in use and the particular question being asked
   - High quality standards are not available
   - Lack of established procedures
   - Collection of known fecal sources
   - Inconsistent levels of matching (typically 80–100%)
   - When and perhaps what to sample (sediments?)

3. What are the common standardization elements for source tracking methods?
   - Follow SOPs available for some of the steps
   - Standards for known samples
     - feces
     - DNA
     - isolates
   - Verification of target organism (biochemical versus genetic approach)
   - Handling of water and fecal samples
   - Reference databases

4. Which source tracking methods are sufficiently standardized? None! Some might be in better shape compared to others (at least some of the steps involved could follow standard procedures).

5. What are the specific needs or research gaps in method standardization for source tracking?
    - Publicly available databases
    - Unify/merge existing databases
    - Review what other researchers are currently doing
    - How to collect and process samples
    - When to collect samples
    - Representative sampling (per station, per watershed)
    - The appropriate number of isolates per sample, station, watershed
    - Validation of genetic methods
    - Standard operating procedures (SOP) for some of the common steps (e.g., PCR protocols)
    - Controls for PCR based methods

The group also discussed the reasons for standardization and approaches for assessing the performance of MST methods. Analytical methods and the results they generate must be able to survive scientific scrutiny. Standardization is necessary to make MST scientifically defensible. It will also be essential for legal defensibility when the results of MST studies inevitably end up in the courtroom.

Benchmarks for standardization include:

- Assessing the performance of methods:
    - Identify known sources
    - Identify mixed sources
    - Identify sources in field samples
    - Assess temporal and spatial stability
    - Evaluate inter-laboratory reproducibility
- Development of national databases
- Development of standard operating procedures
- Meet standards required by monitoring regulations and the legal community

Possible obstacles to the implementation of MST, regardless of the level of standardization are:

- Ethical, medical and legal issues
- No consensus regarding the most appropriate organism(s) to use
- Stakeholder cooperation in obtaining samples
- Current methods may not be adequate
- The lack of universally acceptable standards for each method
- Communication with stakeholders

Recommended research topics included:

1. Develop an understanding of the natural history of fecal indicators in the environment.
2. Assess the spatial and temporal stability of markers.
3. Develop a full characterization of sewage samples.

4. Thoroughly evaluate the performance of current MST methods.
5. Assess the genetic variation in sources being used for each MST method to determine the minimum number needed for accurate identification?
6. Perform a large study encompassing land use, geospatial information systems (GIS) data, hydrological data, and MST to identify hotspots.
7. Determine the relationship between indicator survival outside the host and human health
    - e.g., assess the relative roles of recent contamination versus organisms surviving in sediment.
8. Identify standards for reference materials and sampling methodology.
9. Develop a database for housing and exchanging data from different laboratories.
10. Review published methods to determine their differences and areas of commonality.
11. Develop SOPs for basic techniques that are currently used as components of some methods and are accepted, e.g., basic PCR.
12. Validate the genotypic methods currently in use.
13. Develop a methods comparison project in which a single statistical analysis is used to evaluate the performance of all methods.

Additional issues that need to be addressed include:

- Independent evaluation of commercial laboratories offering MST services.
- Ethical issues regarding human and animal donors of fecal samples.
- Health risk associated with handling of feces

In addition, the MST community can benefit from the experiences of other fields. For example, accreditation is becoming increasingly important for analytical laboratories involved in clinical diagnostics, food and cosmetic testing, forensic analyses, and regulated monitoring. While the MST field is not yet sufficiently mature to be constrained by accreditation criteria, a move in this direction would encourage the development of consensus methods and standardization. Methods for the detection and analysis of chemical contaminants were in a similar condition in the 1980s as the current MST methods. The problem was solved largely by the introduction of performance standards.

4.5 Information Handling

Table 4-8. Information Handling Breakout Group Members.

Tom Edge (group lead)	
James Kern	Mark Borchardt
Christopher Kitts	Jayson Wilbur
Stephen Weisberg	Rick Meinersmann
Katharine Field	

Key issues considered by this breakout group were the options available for data management and maintenance, the advantages and disadvantages of different data analysis methods, and selection of appropriate statistical analysis tools.

Questions Presented to the Breakout Group

1. What are the data management and maintenance needs of source tracking studies?
   - ♦ Standardization of methods to generate data.
   - ♦ Ability and willingness to share data.
   - ♦ Publicly accessible database to evaluate libraries and MST assumptions, e.g., a Genbank for MST.

     However there is neither an immediate need nor consensus on the value of such a national (or international) database, so this would be a long-term goal/option. In lieu of such a database, consideration should be given to utilizing existing libraries on a broader scale with more researcher access. Possibly the most appropriate option at the current time is to merge a number of existing libraries into a single database and ensure that the database is widely accessible.

2. What are feasible options for managing and maintaining source tracking data?
   - ♦ Large public database could be long-term goal but we currently lack standardization, which prevents creation of a useable database.
   - ♦ BioNumerics software (Applied Maths) is widely used for data analysis and management but it is prudent to avoid prescriptive approaches at present. A variety of data analysis and management tools should continue to be used and developed.

3. What are appropriate data analysis methods for source tracking?
   - ♦ The are a variety of data analysis methods currently in use:

     QA/QC – Rarefaction curves.
     Clustering.
     Principal component analysis.
     Discriminant function analysis.
   - ♦ Selection of the most appropriate method depends upon the particular scenario being investigated, the questions being addressed, and the analytical techniques used.
   - ♦ It is important to avoid being too prescriptive at present.

4. What are the research gaps in source tracking data management and analysis?
   - Standards are needed for reporting methods and data.
   - Quality assurance criteria are needed for method evaluations.
   - Since there is a consensus that a toolbox approach is the only realistic way forward for field application of MST, there must be an ability to integrate results from multiple methods. The approaches for such integration are currently limited and not well understood.
   - There needs to be a more thorough evaluation of pattern matching algorithms.
   - Methods for calculating uncertainty (error bars, confidence limits) for MST data.
   - A more in depth discourse is required on appropriate or acceptable library size, including an assessment of the tools that can be used to inform this decision (i.e., what is an appropriate library size? and what *a priori* methods are available to help in answering this question?).

Possible obstacles to the improvement of data analysis and management include a lack of standard methods of data generation, the availability of publicly accessible databases, and the willingness (or unwillingness) of researchers to share data. The comment from this workgroup that you "can't polish a turd" was highly relevant to the topic of the workshop. If a method just does not work (i.e., can not discriminate between fecal samples from various animal hosts), then no amount of improved information handling, data management, standardization, or continued evaluation will make a difference. The method still won't work.

Possible research projects for future funding include:

1. Assessing library quality.
2. Library verification studies with large libraries.
3. Comparison of pattern matching algorithms.
4. Developing QA/QC standards for evaluating methods.
5. Developing approaches for integrating results from multiple MST methods.

4.6 Immediate and Long-Term Research Needs

Table 4-9. Immediate and Long-Term Needs Breakout Group Members.

Ricardo De Leon (group lead)	
Jorge Santo Domingo	Cindy Nakatsu
Kenneth Hyer	William Hamilton
Samuel Myoda	Marion Savill
Jeffrey Ram	Peter Hartel

Questions Presented to the Breakout Group

1. What are the key end users and their information needs?
2. What are their respective research needs?
3. Which ones are short term and which are long term needs?
4. What areas of watershed management benefit from source tracking studies?
5. What decision pathways could source tracking results affect?

The key end users of MST technologies are:

- Regulatory agencies- county, state, regional, U.S. EPA, USDA,
- FDA, CDC, USGS, water control boards.
- Wastewater and drinking water utilities
- Consulting firms
- Parks and recreation departments
- Engineers
- Conservation authorities
- Environmental activist groups
- Farmers
- Scientific organizations
- There is a hierarchy

Additional issues that need to be addressed include:

- Developing approaches for communicating to the public the pros and cons of source tracking.
- Selecting a suite of tools (combination of methods) that can be used for routine field applications in the next three to five years?
- Method improvements to enhance reliability and permit quantification and uncertainty analysis (confidence limits).
- Develop bird contamination markers.
- Interdisciplinary science to resolve many of the outstanding issues is seen as a long-term research need.
- Redundancy and/or confirmatory methods need to be built in to MST approaches. However, based on the current state of the science, it is realized that supposedly confirmatory methods may be confounding.

Important information that is needed in the public domain is reports on the effectiveness of remedial actions that are put in place as a result of MST and the results of MST field studies, even though they do not necessarily report new science. Also, information on the impact on TMDL allocations, site specific standards, discharge allocations, and other permits is required.

Possible obstacles to the successful application of MST are:

- Limited exchange of information.
- Proprietary information.
- Perception of high cost- Both short-term and long-term cost benefit analyses are important in deciding whether to implement MST.
- Case studies are not always novel and therefore rarely considered for publication in peer-reviewed journals.

Potential research topics include:

1. Developing systematic approaches for analysis of MST data.
2. Adopting performance criteria for MST methods and utilizing independent organizations for formalizing methods (i.e., ASTM).
3. Assessing the long-term incentives for continuing MST research and development (i.e., what is the need for MST beyond the TMDL process?).
4. Develop cohesive near-term approaches for MST using the most suitable currently available methods- an interim toolbox approach.
5. There is an immediate need to develop bird-specific markers or a bird-specific method.
6. Method evaluations and studies that encourage interdisciplinary teams.
7. Identify host specific genes.
8. Develop alternative and new chemical methods for source tracking.
9. Greater application of bioinformatics tools to analyze genomic data to increase the pool of potential targets.

CHAPTER 5.0

SUMMARY OF INFORMATION GAPS AND RESEARCH NEEDS

Most of the peer-reviewed literature on MST focuses primarily on developing and evaluating methods. The majority of actual field applications, including some of the most successful case studies, tend to be in the form of non-peer reviewed reports, often distributed by the client or funding organization, or on method developers' websites. There needs to be greater emphasis on the publication of examples of successful application of MST methods. A criticism raised by a number of participants was the lack of peer-reviewed support for some methods and generally inadequate documentation of results. Failure to publish has been referred to as a "scientific crime" (Clapham, 2005) and many consider a scientists work to be wasted without publications. Supporting this sentiment, the view was expressed by workshop participants that all research scientists have an obligation to publish their results, especially if they receive public funding to support their research.

While there should be greater emphasis on publication and dissemination of MST results, there is already a lot of material available in a variety of formats that can make it difficult for potential end-users to make informed decisions. An additional breakout group comprised of workshop observers recommended the establishment of an MST database, providing information on the status and relative costs of methods, availability of contract laboratories, location of MST researchers, links to peer reviewed literature, and copies of non-peer-reviewed agency and other customer reports. This need was echoed by many of the invited workshop participants. Most of this information is already available online, but not necessarily in a single location, as well as in technical journals and newsletters. Many MST researchers maintain websites, some of which include extensive links to a variety of MST resources. There would be relatively little effort involved in converting some of these existing websites into the type of broad MST informational resource envisaged by some workshop participants. Some of the more useful websites that may provide the core for such a database are listed at the end of this report. A possible obstacle to the creation and maintenance of broad informational databases is the level of resources, both time and money, necessary to ensure its completeness and continued relevance.

There are three types of databases or repositories that need to be considered, two being digital only (computer/internet-based) and the third being a physical facility: 1) Centralized informational databases (as described above); 2) A data warehouse; and 3) A physical repository for strains. Database creation and maintenance can be an informal function performed by individual researchers and other interested parties, usually with minimal funding but imparting a considerable burden on the individuals involved. Alternatively, they can be created and maintained by an independent organization, which will usually require considerably higher levels of funding. The open-access advocate BioMed Central (www.biomedcentral.com) maintains a listing of many currently available biologically-related databases, including bioinformatics, infectious diseases, microbiology, genomics, and public health. This organization also provides a database creation and curation service specifically tailored to research communities.

A centralized database of actual MST results from method evaluations, method comparisons, and field studies might contain ribotyping, PFGE, and rep-PCR banding profiles, patterns of antibiotic resistance, or sterol fecalprints. While the most important use of this type of database would be data from known source isolates (library isolates), results from field samples could also be included. Creation of a centralized profile database from scratch would be a significant and costly undertaking, for both start-up and subsequent maintenance. At the Toronto workshop (Chapter 3.0), there was some discussion on the value of a centralized national database, particularly in light of the considerable doubt over the applicability of many MST tools over broad geographic regions.

As an alternative to building a results database from scratch, existing individual researcher or agency maintained databases may be merged. This would increase the total number of known source samples covered by the library, thereby increasing its representativeness and source prediction accuracy. A possible obstacle to merging individual data libraries is that all data in the merged database needs to be generated using the same standardized procedures. It may also be necessary to address issues of data sharing and proprietary information. For example, probably the largest individually maintained library (120,000 *E. coli* ribotype profiles) is maintained by a private company. Libraries also have to be maintained. Since source libraries are not static, they will need to be continually updated to maintain their relevance. An estimate of increasing the library size by 10% every year was provided during the workshop.

The next step beyond a centralized database of MST method results (isolate profiles) is a physical repository of actual isolates. A function of this repository would be to provide a library of standardized, known source strains for evaluation of method accuracy, ensuring that all researchers are using known source isolates of confirmed identity. However, this may need to be a long-term goal since it would require substantial commitment and funding. Nevertheless, creation of a centralized strain library would be an important step in standardizing methods and would probably be essential for meaningful validation studies. The American Type Culture Collection (ATCC) provides a service for storing and distributing specialized culture collections and the Biological Research Collection Program (BRCP), an organization within the National Science Foundation, provides support for biological collection enhancement and management on an annual basis (but the program will not fund creation of a new collection). In addition, the BRCP funds symposia and workshops related to management of biological collections. The International Society for Biological and Environmental Repositories (www.isber.org) also offers a variety of resources to help establish and maintain biological specimen collections.

It was concluded by the majority of workshop participants that it is too early to completely discount any methods. While continuing to evaluate poorly performing methods dilutes research efforts and funding, the timing of a decision to drop a particular method is critical. A method must be allowed to fully develop before such a decision can be made. In addition, methods that do not perform well in isolation may have a role in a toolbox approach. For example, chemical methods appear to have been largely dismissed a few years ago and yet recent work on fecal sterols and fluorimetric detection of optical brighteners indicate that chemical methods should be revisited. In addition, some workshop participants consider the library-based methods to be too involved, expensive, and time-consuming to be practical and should therefore be dropped from future consideration. However, it is important to be careful and not jump to conclusions too early because library-based methods have the potential for the highest level of discriminatory resolution. Therefore, criteria need to be developed that will allow the MST community to make an informed decision about the long-term prospects of individual methods. These criteria should be both quantitative and qualitative; ARCC values, reproducibility, sensitivity, specificity, and the likelihood of success in improving the method with continued research. Of course, no researcher can be forced to drop a method but there may come a point at which the wider MST community decides a method is unworthy of further consideration.

Measures of data confidence need to be developed for all methods. This will allow an assessment of the reliability of MST results. The fact that ARCC values are dependent (up to a point) on the library size (see Figure 1-3) raised concern among some participants, particularly considering that the ARCC is used extensively to evaluate method performance. Techniques used to evaluate the performance of a method should be independent of the method itself.

A questionnaire distributed to workshop attendees and potential end-users asked responders to rank the most important criteria that they would use in selection of source tracking methods for field application. Among a variety of responses, the most frequent criteria were:

- Reproducibility.
- Consensus among experts on the most appropriate method(s).
- Cost of analysis.
- Ability to give absolute answers rather than a qualified “maybe”.
- Standardization.
- Speed of analysis.
- Technical simplicity of the method.

While these are important considerations and apply, or should apply, to all methods (e.g., increased reproducibility and standardization), it is unlikely that a single method will fit all criteria. For example, the cheapest method may not be the most rapid or the most reproducible. Therefore, this mix of competing criteria, with no single method rising above the pack, reinforces the need for a toolbox approach. Also, considering the current state of the science for most of the methods and accepting the majority perception that none of the methods are yet ready for routine and robust use in the field, it may be more appropriate to ask “What can NOT be done?” rather than “What can be done?”

Figure 1-4 in Chapter 1.0 presented the various theoretical levels of animal source discrimination that can potentially be achieved by MST methods. There is still considerable debate over the actual level of discrimination that is possible as well as the level that is necessary to make TMDL and watershed management decisions. In selecting MST methods and method combinations for particular field applications it will often be necessary to strike a balance somewhere along the MST continuum (Figure 5-1). At a utility management level, discrimination between human, livestock, and wildlife sources of contamination will usually be sufficient. Management level discrimination can be achieved with less expensive methods, possibly requiring less subjectivity in interpreting the results, and provide sufficient information for proceeding with the TMDL processes or other remediation measures. A Canadian wastewater perspective (presented at the Toronto workshop) was for discrimination between human and non-human sources and quantification of the relative proportions of each of these contributing sources. Methods with two- or three-way discrimination may also result in fewer isolates being classified as unknown. However, there will be circumstances where three-way discrimination is insufficient and the end-user needs to identify individual types of contributing animal. A variety of discrimination levels are reported (Figure 5-2) but more research is needed to determine the level of discriminatory resolution that can be reliably achieved, and is acceptable, with the various methods and, more importantly, with combinations of methods. In addition, it will be necessary to determine the confidence that can be applied to increasingly complex levels of resolution. The MST field may need increasingly sophisticated methods to test underlying assumptions and source specificity in the laboratory, but then simpler, less costly methods for application by many end-users. Once in the field, it may also be necessary to balance an ideal statistically supported sampling plan against what is practical given funding and time constraints. From a decision-making perspective, it is important that studies are performed to verify that changes in management practice, implemented as a result of MST investigations, are having the desired effect.

5.1 MST Toolbox

There was overall agreement that the most appropriate approach for moving source tracking applications forward is to build an MST toolbox. However, research is needed to determine which methods can be considered mature enough for inclusion in the toolbox and what criteria will be used for tool selection for field application. Maturity should be assessed by the level of standardization that is currently practiced for a technique, or the level that can be readily attained in the near future, and demonstration of performance based on measures of specificity, sensitivity, reproducibility, and accuracy. At a minimum, the toolbox should contain at least one microbiological and one chemical method. Preferably, it should contain multiple diverse and complementary methods. Which particular combination of methods is most suitable for individual field applications will depend on a variety of factors (Figure 5-3).

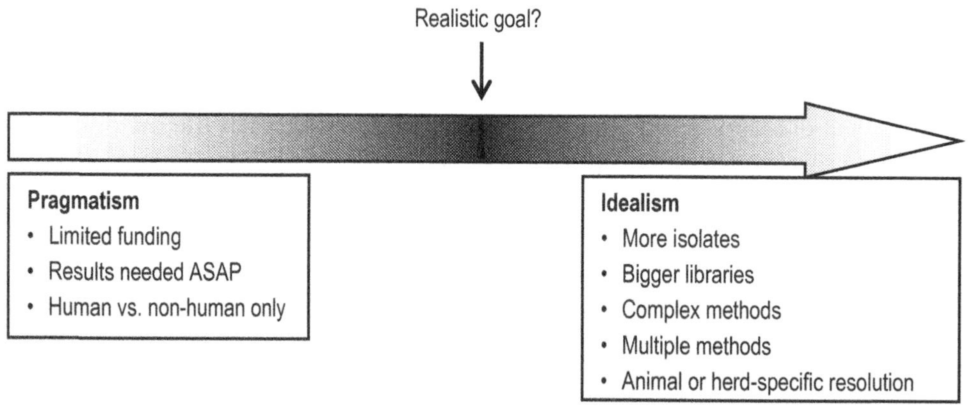

Figure 5-1. The MST Continuum.

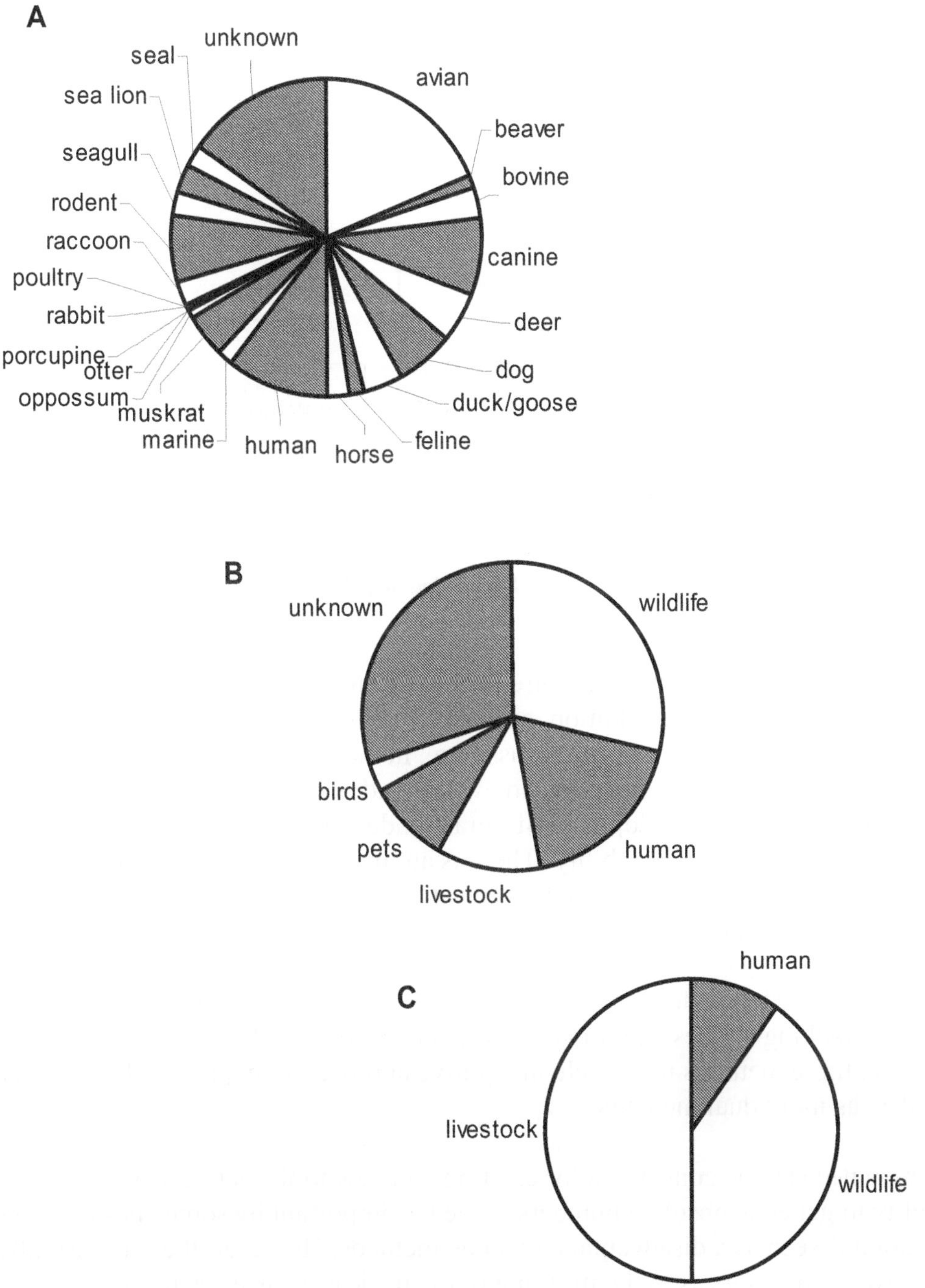

Figure 5-2. Source Discrimination in MST Field Applications Using Ribotyping (A and B) and Antibiotic Resistance Analysis (C). Data sources: Thurston County Public Health Division (www.iehinc.com/mststudies.htm); Whiting-Grant et al., 2003; Graves et al., 2002.

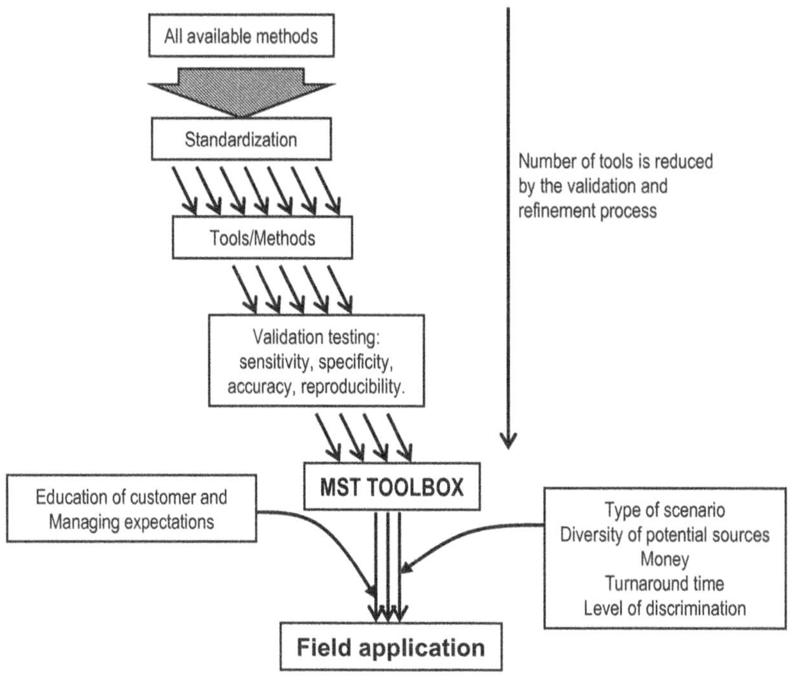

Figure 5-3. Development and Application of MST Toolbox.

All of the methods currently in use require further evaluation along the lines of recent studies to assess their performance. In addition, a suite of markers (existing and new) needs to be developed for each type of host animal or host-class (e.g., human, livestock, and wildlife). Therefore, more research is needed to develop and evaluate additional markers. However, there needs to be a greater focus on the development of library-independent markers and assays since they have the potential for widest applicability. There is an immediate need for bird-specific methods or markers. In addition, chemical methods need a more thorough evaluation and comparison with microbial techniques.

Research is also needed to develop and assess approaches for integrating multiple methods into source tracking studies. An important aspect will be deciding how to analyze the data generated by multiple methods in a single integrative approach (e.g. principal component analysis) rather than as individual methods.

Enumeration needs to be considered for existing and new tools that may be in the toolbox. The ability to generate absolute numbers is seen as important by some end-users and lack of quantification is seen as a disadvantage of some methods. However, there is currently confusion among some end-users as to the meaning of quantification as it relates to MST. Quantification can be achieved on a number of levels:

- How many *E. coli* or enterococci are there in a particular body of water?
- How many copies of the *Bacteroides* 16S rRNA gene are in a sample?
- How much of the contamination can be attributed to each type of animal (i.e., what percentage is human, livestock, seagull, etc.)?

Even when methods are quantitative, there are questions as to what the numbers mean. Are they accurate? Do they reflect the actual proportions of organisms in a watershed or body of water? At best, they may be only pseudo-quantitative, so research is needed to address the interpretation of quantitative methods.

5.2 Library-Based Methods

Even though some researchers feel that library-based methods are too time-consuming and expensive to be practical for long-term use, most participants agreed that no methods should be discarded at this stage in the evolution of MST technologies. However, there are many questions concerning the application of library-based methods:

- How to ensure adequate representation of all sources within the library?
- How to ensure that the library will allow correct classification of unknown isolates?
- What is a statistically appropriate library size?
- How many different host species need to be included in the library?
- How many samples should be collected from each type of source animal?
- How many isolates should be collected from individual animals?
- How large an area can be covered by a single library?

Once a library is constructed, it may only be good for one study. Mutations and genetic drift may mean that new isolates are constantly added to the library to ensure it remains representative. According to some researchers, it may be necessary to add 5–10% new isolates to the library every year. A thorough evaluation needs to be conducted of library-based methods addressing not only source identification accuracy, achievable level of source discrimination, sensitivity, and reproducibility, but also temporal and geographic stability of the methods.

5.3 Standardization

One of the many definitions of "standard" that can be used to describe the need in MST science is: Substantially uniform, well established, and widely recognized as acceptable. Standardization is important for all methods because it: 1) increases the reliability of a method; 2) enables comparisons between different laboratories; 3) allows the method to withstand scientific and legal scrutiny. While it is less important for a single researcher to standardize if there is no intent to compare results with other laboratories or to use the results in any kind of decision making process, standardization is imperative if the results are to survive external inspection and acceptance. Some of the factors in MST analyses that need to be standardized include:

- Organism isolation and identification procedures.

  It is imperative that both library and test isolates are correctly identified and that the procedures used to isolate, culture, and identify field isolates are the same as those used for constructing the library. Therefore, the media, incubation conditions, and identification procedures for *E. coli*, enterococci, and other target organisms should be standardized.

- Measures of data quality.
  Methods for measuring the sensitivity, specificity, accuracy, and variability of MST methods need to be standardized.
- Band matching algorithms.
  For the band matching methods (PFGE, ribotyping, rep-PCR) there is disagreement among researchers over the criteria that are used to determine whether a field isolate matches a library strain. Some analysis methods use a similarity threshold (e.g., 94%) above which all isolates are considered to match. Other methods only accept 100% identity as the measure of similarity between two isolates.
- Data analysis methods.
  Many methods are currently used to score similarity between isolates and the choice of scoring method or statistical analysis tool (e.g., discriminant analysis, principal component analysis, multiple regression, maximum similarity) can have a profound affect on the outcome of MST.
- Antibiotics.
  Different researchers use a variety of different antibiotics at varying concentrations (Table 5-1). There are also a variety of methods for scoring antibiotic resistance or sensitivity, e.g., growth/no growth, minimum inhibitory concentrations, or diffusion zone breakpoint.
- Restriction enzymes.
  Type, concentration, incubation conditions, single or double digests.
- PCR-based analyzes.
  Amplification targets, primer sequences, amplification conditions, selection of polymerase enzyme, standardized controls.
- Electrophoresis conditions for all band matching methods.

5.4 Quality Assurance and Quality Control

Incorporation of QA/QC into MST-related research and field studies is critical to ensure that the results are truly useful, are reproducible, and can withstand scientific and legal scrutiny. However, more information is needed on what elements of QA/QC are necessary for MST. Elements of QA/QC should be applied to all stages of a method or study, including collection of known source fecal samples, preparation of media, isolation and identification of target bacteria, DNA extraction, preparation of antibiotics, amplification reagents, probe preparation, and data recording and analysis. Some of the techniques that are used for MST already have established QA/QC procedures that should be incorporated into an overall strategy for MST. For example, Standard Methods for the Examination of Water and Wastewater (APHA) provides protocols for many of the basic methods and the U.S. EPA provides QA/QC guidance for laboratories performing PCR analyses on environmental samples (U.S. EPA, 2004).

5.5 Target Organisms

The organisms most commonly used for MST are coliforms, *E. coli*, enterococci, and to a lesser extent Bacteroidales. However, fecal coliforms are too diverse as a group in their physiology and ecology to be useful for MST. In addition, issues of population heterogeneity and persistence and growth in the environment limit the utility of *E. coli* and enterococci.

Consequently, there is a need to re-examine the use of existing indicators. A much broader array of target organisms needs to be available for source tracking. Consideration of the potential for survival, persistence, and growth of indicator species in aquatic environments (and particularly in sediments) needs to be an important aspect of selecting alternative organisms.

There were also discussions on the use of sewage as a surrogate for human source samples. While recognizing that obtaining sufficient human fecal samples for method evaluation can be difficult, it was concluded that sewage should not be used as a human standard. There are, however, uses for sewage as a human surrogate, even if only for practicality reasons. It was suggested that whenever possible, pure human samples should be used for assessing the performance of methods, but when practicality dictates, less ideal human-origin samples could be used (Figure 5-4).

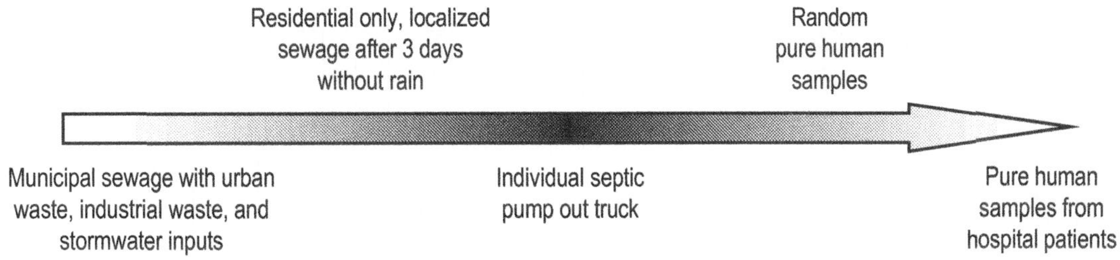

Figure 5-4. The Different Types of Human Source Samples Available for MST Analysis.

Table 5-1. Combinations and Concentrations (μg/mL) of Antibiotics Used in Various ARA Studies.

	Graves et al., 2002	Wiggins et al., 2003	Harwood et al., 2003	Harwood et al., 2003	Harwood et al., 2003	Harwood et al., 2003	Harwood et al., 2003	Ebdon et al., 2004	Sayah et al., 2005*
	Enterococci	Enterococci	*E. coli*	*E. coli*	Fecal coliforms	Enterococci	Fecal strep.	*Enterococci*	*E. coli*
amoxicillin	2.5				4 – 128			4 – 32	
ampicillin				10				4 – 64	10
apramycin								4 – 32	
bacitracin		10 – 100				10 – 100	10 – 100		
cephalothin	10 – 50	10 – 50	10 – 50		8 – 32	10 – 50	10 – 50	8 – 64	30
chloramphenicol					4				30
chlortetracycline	60 – 100	20 – 80	20 – 80	25	20 – 80	20 – 80	20 – 80		
clindamycin								1 – 16	
doxycycline					4				
erythromycin	10 – 50	10 – 50	5			5	10 – 50	8 – 64	
gentamicin		5 – 20	5 – 20		1	5 – 20	5 – 20	4 – 128	10
kanamycin		10 – 50	10 – 50	25	3	10 – 50	10 – 50	4 – 32	
moxalactam			0.2 – 1		0.2				
nalidixic acid				25	3				30
neomycin	40 – 80	10 – 50	10 – 50	50	3	10 – 50	10 – 50	8 – 128	30
nitrofurantoin								4 – 64	300
norfloxacin					0.1				
ofloxacin								2 – 32	5
oxytetracycline	20 – 100	20 – 80	20 – 100	50	20	20 – 100	20 – 80	16 – 128	
penicillin G				75	20 – 200			2 – 16	
polymyxin B			1 – 10						
rifampicin					2 – 16			2 – 32	
streptomycin	40 – 100	20 – 80	20 – 80	12.5	200 – 800	20 – 80	20 – 80	8 – 64	10
sulfathiazole				500				16 – 128	250
tetracycline	10 – 100	10 – 50	10 – 50	25	4 – 64	10 – 50	10 – 50	2 – 64	30
trimethoprim					0.25 – 1				
trimethoprim/ sulfamethoxazole			0.2 – 5		5				1 – 24
tylosin								2 – 32	
vancomycin	2.5	5 – 30				2.5 – 5	5 – 30	4 – 32	

* Disk diffusion method so quantity is amount of antibiotic per disk rather than concentration.

CHAPTER 6.0

CONCLUSIONS AND RECOMMENDATIONS

6.1 Overview

There was vigorous discussion amongst workshop participants on many of the issues relating to FST. While there was dispute on some aspects, there were also many areas of agreement. This chapter summarizes the overall outcome of the workshop and makes recommendations for the future of FST, based on a consensus of the most important issues. The workshop organizers maintained neutrality throughout the meeting and in summarizing the outcome of the workshop. The general conclusion from the workshop was:

> "There is almost certainly a role for FST in the wastewater and drinking water industries if it is part of an overall strategy that includes multiple congruent methods and traditional inquiries (land use, sanitary, and wildlife surveys). In addition, the geographic scale of the study area should be relatively small, and defined criteria of standardization, QA/QC, and measures of data quality must be established".

There are reports of the successful application of MST technologies in relatively small-scale field studies with two-, three-, and four-way source discrimination. However, without further improvements in method accuracy and robustness, it is unlikely that any of the proposed methods can consistently identify fecal contamination at the level of individual host animals. Numerous methods have been proposed for FST and there has been considerable debate over the most appropriate techniques. However, it is becoming increasingly apparent that no single technique will be suitable in all situations and that a "toolbox" approach involving multiple methods will be required. This realization is not new; Sinton et al. (1998) reached the same conclusion and stated that "statistical analysis of a basket of microbial and chemical determinants offer the possibility of identifying and apportioning human and animal faecal inputs to natural waters". Developing and refining a variety of the most appropriate methods is warranted to ensure a sufficiently large suite of techniques is available for the toolbox for different applications. The toolbox components in a particular study are likely to vary depending on the:

- Geographical location, limnology, hydrology, and physical characteristics of the watershed or body of water under investigation.
- Diversity and number of potential source(s) of contamination.
- Required or acceptable level of discriminatory resolution.
- Required level of certainty in the results.
- Level of funding available.
- Urgency to generate results.

One of the primary drivers for moving FST technologies out of a research setting and into "real-world" applications in the U.S. is TMDL requirements under the Clean Water Act. This is considered a premature move by many researchers. Nevertheless, it means that MST is not simply an academic exercise. For example, MST is an integral part of the State of Virginia's TMDL process and the cost of MST over the next five years is estimated at 10% of the state's total $10.7 million to develop the necessary TMDLs through 2010 (Virginia Department of Environmental Quality, February 2005). There are therefore legal, political, and financial consequences of MST-derived conclusions and remediation programs. Repeatability, variability, reproducibility, and reliability are all important because researchers may find themselves in court defending MST methodology when they are used to inform the TMDL process. Consequently, all aspects of MST methods must be scientifically sound and legally defensible.

For water and wastewater utilities and other end-users attempting to identify sources of fecal contamination using an FST approach, currently available options include:

- Developing in-house capabilities (this favors the simpler and less costly cultural approaches).
- Collecting samples/isolates and then paying a commercial service to perform the analyses.
- Contracting out for the entire investigation (likely to be the most expensive approach, depending on the techniques).
- Collaborating with individual method developers in academic institutions.

Whichever approach is adopted, utilities need to have a clear understanding of the questions they are asking, the expected level of resolution, an acceptable level of ambiguity, how to interpret results, what kind of results to expect, and the costs involved. Issues of quality assurance, method reproducibility, and statistical robustness should also be addressed. FST should never be the only tool used to determine the source(s) of fecal contamination. Source tracking programs also need to include studies in the watershed of:

- Agricultural land use management practices.
- Surveys of sewage and septic systems.
- Stream/river flow patterns.
- Seasonal run-off patterns.
- Sewage treatment practices.
- Wildlife surveys.

Irrespective of source tracking studies, it is also important to take sensible remediation measures, which may in the long-term be a better investment of limited resources. Such actions may include: keeping cattle and other livestock out of drinking water reservoirs by constructing or repairing fences; deterring large-scale wildlife flocking and herding activities in and around reservoirs; locating and repairing damaged sewage lines; and improving land management practices in watersheds. There are examples of MST being used to guide remediation programs. For example, ribotyping *E. coli* isolates from the Webhannet River watershed apportioned sources of contamination as 29% wildlife, 18% human, 11% livestock, 9% pets, 3% birds, and 30% unknown (Whiting-Grant et al., 2003). Following the study, recommendations were developed to reduce fecal contamination from each of these sources. However, all of the recommendations were sensible measures that could have been recommended without first conducting an MST study: fix malfunctioning septic systems, repair leaking sewer pipes, reduce attraction of wildlife, improve handling of livestock waste, and promote appropriate handling of pet waste. None of the recommendations was dependent on the source tracking study. In fact, since the single largest source was unknown, the study served to demonstrate a weakness of the method.

While many issues were discussed during the workshop, some topics became recurring themes over the three day period and are therefore seen as some of the most critical areas for consideration:

- Cost of analysis: Practical considerations often result in the simplest and cheapest method being desirable, regardless of its actual performance.
- Availability of funding.
- Techniques must be legally defensible.
- The role of sediments needs to be considered.
- More evaluation of methods is required.
- There is a need for new methods and targets.
- Standardization and QA/QC are critical.
- Different combinations of methods need to be evaluated.
- Information needs to be provided to customers to maintain realistic expectations and avoid disappointment.

6.2 Conclusions

In recent years, most researchers involved in MST have coalesced into a group that is prepared to cooperate and collaborate, participate in inter-laboratory methods evaluations, and share data. This provides the necessary framework for future method development and improvement of MST approaches. Continued communication within the research community and improved communication between researchers and end-users is vital for the future success of MST technologies. The workshop produced the following conclusions regarding critical issues that need to be addressed if MST is to continue moving forward:

- Recent method comparison studies have given researchers a "reality check" on the performance of MST methods. It is now necessary to communicate this to customers in terms of realistic expectations and limitations. A significant lesson learned from this and other recent MST workshops is the recognition of the need to manage the expectations of technology end-users. In some instances, utility managers, administrators, regulatory agencies, policymakers, and other end-users may have unrealistic expectations about the outcome of MST programs. This is due to a combination of techniques being "oversold" by some method developers, and misinterpretation by end-users. While potentially eroding public confidence in MST, the reality check provided by recent comparisons has in fact increased the credibility of the field since researchers are recognizing and publishing the limitations of the methods. Conducting science in the public eye does not always cast the results in a favorable light. Therefore, in addition to conducting the best possible science to further develop and evaluate MST technologies, there is a need for public outreach so that potential customers know what can, and cannot, be achieved by MST. It is important that researchers remain objective in their evaluation of methods and strike an appropriate balance between being too critical and dismissive of methods versus being too optimistic and over-selling MST tools.

- It is obvious that the "silver bullet" ideal method will probably not be developed. One size/target/method will not fit all application scenarios. Therefore, a toolbox approach is most appropriate for field applications. However, this necessitates rigorously tested, validated, and standardized methods to go in the toolbox. Arbitrary comparisons (e.g., genotypic versus phenotypic, library-dependent versus independent, culture-based versus culture-independent, and chemical versus microbiological) are not useful. Most of the methods have unique merits that may lend themselves to particular situations. However, a distinct advantage of the toolbox approach is that congruence between different methods increases confidence in the results.
- Although there are reports of discrimination at the level of individual host species, recent comparison studies cast doubt on the ability of any of the techniques to achieve this resolution on a routine and consistent basis. However, it can be concluded that some MST methods can reliably discriminate between human and non-human sources of contamination.
- More inter-laboratory comparison studies and evaluation of methods are needed. While evaluation of individual methods is still necessary to assess their source prediction accuracy, reproducibility, sensitivity, and specificity, the primary focus of future investigations should be comparisons between various combinations of methods.
- There is an immediate need for standardization because many comparisons show poor reproducibility among investigators. However, it is currently not clear which methods are sufficiently mature to be standardized. There was a consensus that standardization is one of the most important elements for MST in the near future. It is probably less important to be concerned about data analysis and management, statistics, and study design. If methods do not work, or do not work consistently because of lack of standardization, then no amount of discussion on these other topics will make any difference. Once standardized, published or otherwise publicly available SOPs must be adopted and followed by anyone performing a particular technique, if they expect their results to be given serious consideration by the wider MST and end-user communities. Standardization is necessary if the techniques and their results are to withstand scientific scrutiny and legal challenges. Some methods are considered more mature and possibly ready for standardization while others are still experimental or research grade tools requiring further refinement, optimization, and evaluation.
- Incorporation of QA/QC measures into all aspects of MST research and application is critical. If methods are going to be driven out of the research laboratory into real-world applications before they are sufficiently mature, then incorporation of extensive QA/QC measures will be necessary.
- Performance evaluation criteria need to be developed so that all researchers are using the same approach when developing, evaluating, and comparing methods. In addition, it may be necessary to establish a mechanism for independent oversight of method evaluations or for independent testing of methods. For example, organizations such as ASTM International and the U.S. EPA's Environmental Technology Verification

(ETV) provide avenues for independent testing and standardization of analytical methods.

- Continued development of new methods is necessary. Such methods may include multi-target microarrays, eukaryotic mitochondrial DNA, new host-specific organisms or molecular markers in specific organisms. Metagenomics should also be investigated as a process for target selection. In addition, some existing chemical methods, particularly fecal sterols, are worth revisiting.
- Improved documentation of all aspects of MST research and field applications is necessary. This includes more case study publications, particularly increased publication of success stories, more detailed method descriptions, documentation of QA/QC measures, and the steps taken to assess data quality.
- Since some methods may already be, or will soon become, mature enough to progress beyond the research laboratory, it will be necessary to develop a mechanism for achieving this transition. For example, what level of standardization is necessary, how many comparison studies need to be conducted, what are acceptable levels of source prediction accuracy, sensitivity, and specificity, and what levels of intra- and inter-laboratory reproducibility are required before a tool or suite of tools can be moved forward to field applications? Once these criteria are met, a pilot-scale evaluation should be performed prior to full-scale application to ensure that the method is still valid outside of the research setting.
- There needs to be a stronger design element in MST field application studies that takes into account appropriate sampling regimes, including sample site locations, number and frequency of samples, number of isolates, types and number of potential source animals, number of isolates per animal, and number of replicates per isolate.
- Selecting appropriate data analysis tools is seen as a very important issue since there is growing concern about obtaining different MST answers depending on the method of data analysis that was used. There is no consensus on using 100% identity or <100% similarity as matching parameters in band-matching methods. Nevertheless, it is necessary to determine the similarity threshold for band-matching methods such as rep-PCR, ribotyping, and PFGE. This will involve performing multiple replicate analyses on a single isolate, with and without successive sub-culturing and then analyzing the results using a variety of band matching algorithms and clustering tools. The lowest similarity score is then adopted as the similarity threshold for that particular method.
- Apart from the information sharing and technical improvements that are needed for MST methods, there also needs to be improved awareness of what the results mean in respect to public health and how they related to the physical, chemical, and biological characteristics of a watershed or body of water. A greater understanding is needed of the relationship between watershed land use and the sources of bacteria and/or indicators in streams. Research also needs to be conducted on the role played by sediments in "storing and disseminating" fecal signatures and their potential confounding effect on MST.
- Mechanisms should be established to increase sharing of data, possibly through the creation of publicly available databases. In addition, a comprehensive FST database would provide a great benefit to the entire end-user community. Information to include in such a database includes the status, relative costs, advantages, and limitations of

various methods, availability of contract laboratories, location of MST researchers, and links to peer reviewed literature

- MST studies should focus on locations where remediation efforts are planned so that results prior to, and following remediation can be compared. These comparisons should be both short-term (i.e., immediately following intervention) and long-term. Whatever the long-term objective or driver behind the application of source tracking, MST should only be considered successful if the resulting remediation measures are successful (i.e., contamination is reduced).
- Funding of MST-related research and of field applications is a critical issue. There must be sufficient funding on the research side to allow implementation of standardization, QA/QC, rigorous testing, method comparisons, and development of new methods. In addition, field studies must be adequately funded to allow sufficient samples to be analyzed with the most appropriate methods. It is recognized that study scale depends on available funding which relates to the potential public health impact. Consequently, outreach to research funding organizations, government agencies, the wastewater and drinking water industries, and other end-users, is required to promote adequate resources.

6.3 Recommendations

Some of the recommendations that were developed during the workshop relate to broad policy issues such as increased funding and improved communication. Others deal with specific areas where the science of MST needs to progress, some of which will require long-term programmatic funding efforts while others may be accomplished in a single project.

6.3.1 Policy Recommendations

- There is currently the potential for conducting small-scale MST field studies using methods that can reliably distinguish human from non-human sources of contamination. However, many areas within the MST field need continued research. Consequently, considering the current state of the science, the wastewater and drinking water industries should exercise caution when contemplating the use of MST methods. The technologies are not yet ready to be used reliably and consistently for supporting watershed management or TMDL-related decision making processes. However, despite this lack of maturity with the methods, they are likely to be used with increasing frequency over the next few years. Therefore, a useful role for interested organizations such as WERF in the ongoing development of MST would be to sponsor laboratory and field-based validations and inter-laboratory comparisons of various method combinations (the toolbox approach) while promoting the development of standardized methods and QA/QC measures.
- Creation of a long-term, multi-agency program for funding MST research is necessary to ensure sufficient scientific progress. While MST studies, method evaluations and comparisons have been funded by a variety of agencies, continued progress will be accelerated with a coordinated, programmatic multi-year funding effort. Therefore, a mechanism should be established to facilitate cooperative funding between agencies with a stake in the development and successful application of MST. Such agencies might include WERF, U.S. Environmental Protection Agency, Awwa Research

Foundation (AwwaRF), Environment Canada, National Oceanic and Atmospheric Administration, U.S. Geological Survey, and U.S. Dept. of Agriculture. Funding partnerships are necessary to support larger more in-depth studies and to allow the inevitable follow-up studies. It is suggested that during the development of requests for proposals (RFP) related to MST studies, pre-RFPs are sent to prospective funding partners to gage the interest in co-funding.

- An end-user workshop or meeting should be organized aimed at instilling realistic expectations. Specifically, end-users need to be made aware of what is, and what is not possible with current techniques. This meeting would not present a detailed review of methods or comparison studies. Rather, it would present an overview of the advantages and disadvantages of different methods, limitations of the technology, the types of questions that may be addressed by MST technologies, and the type of auxiliary information that is necessary for successful application of MST. Ideally, this workshop would be facilitated by individuals with no stake in any particular technique. A potential venue for such a meeting would be the Water Environment Foundation's annual WEFTEC conference. Alternatively, a white paper based on an update of the information in this report may be an effective means of communicating with MST end-users. Once initiated, this communication should continue so that end-users are provided with regular updates on the progress of the science as well as successes and problems arising with field applications.
- Since there will inevitably be MST workshops, seminars, and meetings in the future, coordination between the organizers of these events should be encouraged so that each new gathering of experts builds upon the experiences and lessons learned from earlier efforts, rather than duplicating material and revisiting the same issues.

6.3.2 Research Recommendations

- In a move towards a toolbox approach, there is a need to conduct comparison studies designed to evaluate different suites of methods to determine those that best meet accuracy, sensitivity, and reproducibility criteria. Inter-laboratory comparisons of various method combinations incorporating standardization, QA/QC, and *a priori* criteria for assessing method performance, are necessary. These studies should also assess the different statistical methods that could be used to analyze the integrated dataset that will result from the individual methods within the combination (e.g. principal component analysis).
- It is necessary to determine the most appropriate approach for standardizing methods and incorporating elements of QA/QC into MST studies. Criteria for acceptable performance of a method must be developed along with a strategy for testing these criteria (e.g., how many blind samples are needed?). Submission of methods for independent evaluation using the predetermined criteria and testing strategy is recommended.
- Field studies using a combination of at least two of the best available methods before and after remediation activities will help to determine whether remediation efforts are successful and also if FST methods can detect (and measure) the effect of remediation.
- More specific research is necessary for many of the methods, including:
  - Host-specific gene markers show promise as a non-library, culture-independent approach to MST. However, a more thorough evaluation of the methods is

necessary. The geographical specificity and inter-laboratory reproducibility of the methods needs to be evaluated along with an assessment of the level of source discrimination that is feasible. Quantification also needs to be assessed.
    - More rigorous testing of library-based methods is required, including developing tools that can be used to determine the appropriate library size for any particular study.
    - Spatial and temporal stability of all markers needs to be investigated more thoroughly.
    - Further investigation of chemical methods is necessary to address questions about persistence, biodegradability, and correlation to microbial indicators.
    - Develop a suite of standardized controls for all PCR-based methods, e.g., positive and negative controls, reagent controls, extraction and matrix inhibition controls.
- The fate and transport of fecal indicators in the environment should be investigated, including assessing the short-term and long-term environmental persistence, ecology, and population genetics of all target organisms and markers. Projects on this issue should also investigate the role of sediments in the fate (death, growth, resuspension, or predation) of indicator organisms after leaving their respective host animals.
- A study is needed to assess the feasibility of merging existing databases for library-based methods and making them publicly available.
- Research is needed to discover alternatives to *E. coli* and enterococci as MST target organisms since intra-species heterogeneity, growth and persistence in the environment, and poorly defined host specificity may limit the utility of these bacteria. Linked to this would be an assessment of the genetic variability of target organisms to allow an estimation of the minimum number of isolates that would be needed for accurate identification.

6.3.3 Recommendations for WERF

The following prioritized recommendations highlight a potential role for WERF in advancing the application of MST technologies by ensuring the highest quality practical science while maintaining realistic expectations:

1. Sponsor inter-laboratory comparisons and field-based trials of various method combinations (the toolbox approach) while promoting the development of standardized methods and QA/QC measures.
2. Fund research that leads to the development of a toolbox approach that can accurately, reliably, and consistently discriminate between human and non-human contamination. This toolbox must be supported by extensive evaluation and pilot-scale field trials, inter-laboratory comparison studies demonstrating high reproducibility, and independent validation.
3. Partner with other organizations in developing a program for pragmatic and defensible applications of MST.
4. Sponsor public outreach events aimed at establishing realistic expectations for potential MST customers.

CHAPTER 7.0

SOURCE TRACKING RESOURCES

7.1 Recent MST Workshops and Symposia

American Water Works Association: Microbial Source Tracking. November 2005, Quebec City, Canada. *Paul A. Rochelle and George Di Giovanni.*

American Society for Microbiology: Microbial Source Tracking Using Indicator Organisms. June 2005, Atlanta, Georgia. *Jill Stewart.*

Environment Canada: Microbial Source Tracking in Aquatic Ecosystems: The State of the Science and an Assessment of Needs. March 2005, Toronto, Ontario. *Tom Edge and Karl Schaefer.*

WERF: Workshop on Microbial Source Tracking in Water. February 2005, San Antonio, Texas. *Paul A. Rochelle and Ricardo De Leon.*

U.S. Department of Agriculture-CSREES: Pathogens and TMDLs- National Water Quality Conference. February 2005, San Diego, California.

American Water Works Association: Pollutant Source Identification Technical Session- Source Water Protection Symposium. January 2005, Palm Beach Gardens, Florida.

U.S. Environmental Protection Agency Gulf of Mexico Program and University of Southern Mississippi: Bacterial Source Tracking Workshop. November 2004, Biloxi, Mississippi. *Rudolph Ellender.*

European Institute of Health and Medical Sciences: Conference on Source Tracing of Water Pollution. January 2004, Guilford, England.

U.K. Department for Environment, Food and Rural Affairs: Approaches to the Detection and Identification of Faecal Sewage Contamination in Coastal Waters. September 2003, Oban, Scotland.

The Society for Environmental Toxicology and Chemistry 24th Annual Meeting: Fecal Source Tracking. November 2003, Austin, Texas.

U.S. Environmental Protection Agency: Microbial Source Tracking Seminar. July 2003, Seattle, Washington.

U.S. Environmental Protection Agency: Workshop on Molecular Methods for Water Quality Monitoring. May 2003, Philadelphia, Pennsylvania.

U.S. Environmental Protection Agency: Workshop on Microbial Source Tracking. February 2002, Irvine, California. *Steven Weisberg.*

7.2 Selected MST Websites

http://water.usgs.gov/owq/microbial.html

http://sun.science.wayne.edu/~jram/MGLPF-MSTProject.htm

http://soils1.cses.vt.edu/ch/biol_4684/bst/BST.html

http://lakes.chebucto.org/H-2/bst.html

http://www.ub.es/microbiologia/TOFPSW_archives/TOFPSW1.html

http://www.cropsoil.uga.edu/mst/

http://www.iehinc.com/mstindex.htm

http://www.sourcemolecular.com/

REFERENCES

Anderson, K.L., Whitlock, J.E., and Harwood, V.J. 2005. Persistence and differential survival of fecal indicator bacteria in subtropical waters and sediments. Appl. Environ. Microbiol. 71:3041-3048.

Avery, S.M., Liebana, E., Reid, C.–A., Woodward, M.J., and Buncic, S. 2002. Combined use of two genetic fingerprinting methods, pulsed-field gel electrophoresis and ribotyping, for characterization of *Escherichia coli* O157 isolates from food animals, retail meats, and cases of human disease. J. Clin. Microbiol. 40:2806-2812.

Bernhard, A.E. and Field, K.G. 2000. A PCR assay to discriminate human and ruminant feces on the basis of host differences in *Bacteroides-Prevotella* genes encoding 16S rRNA. Appl. Environ. Microbiol. 66:4571-4574.

Bernhard, A.E., Goyard, T., Simonich, M., and Field, K.G. 2003. A rapid method for identifying fecal pollution sources in coastal waters. Water Res. 37:909-913.

Berthiaume, F., Brousseau, R., Lemarchand, K., Masson, L., Maynard, C., Payment, P., Bayardelle, P., and Harel, J. 2004. Application of DNA Microarray Technology for Wastewater Analysis. Water Environment Research Foundation, Alexandria, VA.

Blanch, A.R., Belanche-Munoz, L., Bonjoch, X., Ebdon, J., Gantzer, C., Lucena, F., Ottoson, J., Kourtis, C., Iversen, A., Kuhn, I., Muniesa, L.M.M., Schwartzbrod, J., Skraber, S., Papageorgiou, G., Taylor, H.D., Walllis, J., and Jofre, J. 2004. Tracking the origin of faecal pollution in surface water: An ongoing project within the European Union research program. J. Wat. Health. 2:249-260.

Bonjoch, X., Balleste, E., and Blanch, A.R. 2004. Multiplex PCR with 16S rRNA gene-targeted primers of *Bifidobacterium* spp. to identify sources of fecal pollution. Appl. Environ. Microbiol. 70:3171-3175.

Booth, A.M., Hagedorn, C., Graves, A.K., Hagedorn, S.C., and Mentz, K.H. 2003. Sources of fecal pollution in Virginia's Blackwater river. J. Environ. Engineer. 129:547-552.

Booth, J. and Brion, G.M. 2004. The utility of the AC/TC ratio for watershed management: A case study. Wat. Sci. Technol. 50:199-203.

Carson, C.A., Shear, B.L., Ellershiek, M.R., and Schnell, J.D. 2003. Comparison of ribotyping and repetitive extragenic palindromic-PCR for identification of fecal *Escherichia coli* from humans and animals. Appl. Environ. Microbiol. 69:1836-1839.

Chao, A. 1984. Non-parametric estimation of the number of classes in a population. Scan. J. Stat. 11:265-270.

Chaperon, C.D., Ballester, N.A., Fontaine, J.H., Frades, C.N., and Margolin, A.B. 2000. Detection of astroviruses, enteroviruses, and adenovirus types 40 and 41 in surface waters collected and evaluated by the information collection rule and an integrated cell culture-nested PCR procedure. Appl. Environ. Microbiol. 66:2520-2525.

Clapham, P. 2005. Publish or perish. BioScience 55:390-391.

Cole, D., Long, S.C., and Sobsey, M.D. 2003. Evaluation of F+ RNA and DNA coliphages as source-specific indicators of fecal contamination in surface waters. Appl. Environ. Microbiol. 69:6507-6514.

Dick, L.K., Bernhard, A.E., Brodeur, T.J., Santo Domingo, J.W., Simpson, J.M., Walters, S.P., and Field, K.G. 2005a. Host distribution of uncultivated fecal *Bacteroidales* bacteria reveal genetic markers for fecal source identification. Appl. Environ. Microbiol. 71:3184-3191.

Dick, L.K., Simonich, M.T., and Field, K.G. 2005b. Microplate subtractive hybridization to enrich for *Bacteroidales* genetic markers for fecal source identification. Appl. Environ. Microbiol. 71:3179-3183.

Dombek, P.E., Johnson, L.K., Zimmerley, S.T., and Sadowsky, M.J. 2000. Use of repetitive DNA sequences and the PCR to differentiate *Escherichia coli* isolates from human and animal sources. Appl. Environ. Microbiol. 66:2572-2577.

Ebdon, J.E., Wallis, J.L., and Taylor, H.D. 2004. A simplified low-cost approach to antibiotic resistance profiling for faecal source tracking. Wat. Sci. Technol. 50:185-191.

Edwards, D.D., McFeters, G.A., and Venkatesan, M.I. 1998. Distribution of *Clostridium perfringens* and fecal sterols in a benthic coastal marine environment influenced by the sewage outfall from McMurdo Station. Appl. Environ. Microbiol. 41:1139-1143.

Elhmmali, M.M., Roberts, D.J., and Evershed, R.P. 1997. Bile acids as a new class of sewage indicator. Environ. Sci. Technol. 31:3663-3668.

Elhmmali, M.M., Roberts, D.J., and Evershed, R.P. 2000. Combined analysis of bile acids and sterols/stanols from riverine particulates to assess sewage discharges and other fecal sources. Environ. Sci. Technol. 34:39-46.

Farnleitner, A.H., Kreuzinger, N., Kavka, G.G., Grillenberger, S., Rath, J., and Mach, R.L. 2000. Simultaneous detection and differentiation of *Escherichia coli* populations from environmental freshwaters by means of sequence variations in a fragment of the β-D-glucuronidase gene. Appl. Environ. Microbiol. 66:1340-1346.

Ferguson, D.M., Moore, D.F., Getrich, M.A., and Zhowandai, M.H. 2005. Enumeration and speciation of enterococci found in marine and intertidal sediments and coastal water in southern California. J. Appl. Microbiol. 99:598-608.

Field, K.G. 2004. Faecal source identification. In: Waterborne Zoonoses: Identification, Causes and Control, pp. 349-366. (Eds. Cotruvo, J.A., Dufour, A., Rees, G., Bartram, J., Carr, R., Cliver, D.O., Craun, C.F., Fayer, R., and Gannon, V.P.J.), IWA Publishing, London.

Field, K.G., Chern, E.C., Dick, L.K., Fuhrman, J., Griffith, J., Holden, P.A., LaMontagne, M.G., Le, J., Olson, B., Simonich, M.T. 2003. A comparative study of culture-independent, library-independent genotypic methods of fecal source tracking. J. Wat. Health 1:181-193.

Fong, T.T., Griffin, D.W., and Lipp, E.K. 2005. Molecular assays for targeting human and bovine enteric viruses in coastal waters and their application for library-independent source tracking. Appl. Environ. Microbiol. 71:2070-2078.

Gilpin, B., James, T., Nourozi, F., Saunders, D., Scholes, P., and Savill, M. 2003. The use of chemical and molecular microbial indicators for faecal source identification. Wat. Sci. Technol. 47:39-43.

Graves, A.K., Hagedorn, C., Teetor, A., Mahal, M., Booth, A.M., and Reneau, Jr., R.B. 2002. Antibiotic resistance profiles to determine sources of fecal contamination in a rural Virginia watershed. J. Environ. Qual. 31:1300-1308.

Gordon, D.M. 2001. Geographical structure and host specificity in bacteria and the implications for tracing the source of coliform contamination. Microbiology 147:1079-1085.

Gordon, D.M. and Cowling, A. 2003. The distribution and genetic structure of *Escherichia coli* in Australian vertebrates: host and geographic effects. Microbiology 149:3575-3586.

Gordon, D.M. and Lee, J. 1999. The genetic structure of enteric bacteria from Australian mammals. Microbiology 145:2673-2682.

Gordon, D.M., Bauer, S., and Johnson, J.R. 2002. The genetic structure of *Escherichia coli* populations in primary and secondary habitats. Microbiology 148:1513-1522.

Gregor, J., Garrett, N., Gilpin, B., Randall, C., and Saunders, D. 2002. Use of classification and regression tree (CART) analysis with chemical faecal indicators to determine sources of contamination. N. Z. J. Mar. Freshwater Res. 36:387-398.

Griffin, D.W., Stokes, R., Rose, J.B., and Paul, J.H. 2000. Bacterial indicator occurrences and the use of an F-specific RNA coliphage assay to identify fecal sources in Homosassa springs, Florida. Microb. Ecol. 39:56-64.

Griffith, J.F., Weisberg, S.B., and McGee, C.D. 2003. Evaluation of microbial source tracking methods using mixed fecal sources in aqueous test samples. J. Water Health 1:141-151.

Hagedorn, C., Robinson, S.L., Filtz, J.R., Grubs, S.M., Angier, T.A., and Reneau, R.B. 1999. Determining sources of fecal pollution in a rural Virginia watershed with antibiotic resistance patterns in fecal streptococci. Appl. Environ. Microbiol. 65:5522-5531.

Hagedorn, C., Crozier, J.B., Mentz, K.A., Booth, A.M., Graves, A.K., Nelson, N.J., and Reneau, R.B. 2003. Carbon source utilization profiles as a method to identify sources of fecal pollution in water. J. Appl. Microbiol. 94:792-799.

Hane, B.G., Jager, K., and Drexler, H. 1993. The Pearson product-moment correlation coefficient is better suited for identification of DNA fingerprinting profiles than band matching algorithms. Electrophoresis 14:967-972.

Hartel, P.G., Summer, J.D., Hill, J.L., Collins, J.V., Entry, J.A., and Segars, W.I. 2002. Geographic variability of *Escherichia coli* ribotypes from animals in Idaho and Georgia. J. Environ. Qual. 31:1273-1278.

Harwood, V.J., Whitlock, J., and Withington, V. 2000. Classification of antibiotic resistance patterns of indicator bacteria by discriminant analysis: Use in predicting the source of fecal contamination in subtropical waters. Appl. Environ. Microbiol. 66:3698-3704.

Harwood, V.J., Wiggins, B., Hagedorn, C., Ellender, R.D., Gooch, J., Kern, J., Samadpour, M., Chapman, A.C.H., Robinson, B.J., Thompson, B.C. 2003. Phenotypic library-based microbial source tracking methods: Efficacy in the California collaborative study. J. Wat. Health 1:153-166.

Hsu, F.–C., Shieh, Y.–S., van Duin, J., Beekwilder, M.J., and Sobsey, M.D. 1995. Genotyping male-specific RNA coliphages by hybridization with oligonucleotide probes. Appl. Environ. Microbiol. 61:3960-3966.

Hulton, C.S.J., Higgins, C.F., and Sharp, P.M. 1991. ERIC sequences: a novel family of repetitive elements in the genome of *Escherichia coli*, *Salmonella typhimurium* and other enteric bacteria. Mol. Microbiol. 5:825-834.

Isobe, K.O., Tarao, M., Chiem, N.H., Minh, L.Y., and Takada, H. 2004. Effect of environmental factors on the relationship between concentrations of coprostanol and fecal indicator bacteria in tropical (Mekong Delta) and temperate (Tokyo) freshwaters. Appl. Environ. Microbiol. 70:814-821.

Jenkins, M.B., Hartel, P.G., Olexa, T.J., and Stuedemann, J.A. 2003. Putative temporal variability of *Escherichia coli* ribotypes from yearling steers. J. Environ. Qual. 32:305-309.

Jimenez-Clavero, M.A., Fernandez, C., Ortiz, J.A., Pro, J., Carbonell, G., Tarazona, J.V., Roblas, N., and Ley, V. 2003. Teschoviruses as indicators of porcine fecal contamination in surface water. Appl. Environ. Microbiol. 69:6311-6315.

Jimenez-Clavero, M.A., Escribano-Romero, E., Mansilla, C., Gomez, N., Cordoba, L., Roblas, N., Ponz, F., Ley, V., and Saiz, J.-C. 2005. Survey of bovine enteroviruses in biological and environmental samples by a highly sensitive real-time reverse transcription-PCR. Appl. Environ. Microbiol. 71:3536-3543.

Johnson, J.R. and O'Bryan, T.T. 2000. Improved repetitive-element PCR fingerprinting for resolving pathogenic and nonpathogenic phylogenetic groups with *Escherichia coli*. Clin. Diag. Lab. Immunol. 7:265-273.

Johnson, L.K., Brown, M.B., Carruthers, E.A., Ferguson, J.A., Dombek, P.E., and Sadowsky, M.J. 2004. Sample size, library composition, and genotypic diversity among natural populations of *Escherichia coli* from different animals influence accuracy of determining sources of fecal pollution. Appl. Environ. Microbiol. 70:4478-4485.

Khatib, L., Tsai, Y.-L., and Olson, B.H. 2002. A biomarker for the identification of cattle pollution in water using the LTIIa toxin gene from enterotoxigenic *E. coli*. Appl. Microbiol. Biotechnol. 59:97-104.

Kinzelman, J., McLellan, S.L., Daniels, A.D., Cashin, S., Singh, A., Gradus, S., and Bagley, R. 2004. Non-point source pollution: Determination of replication versus persistence of *Escherichia coli* in surface water and sediments with correlation of levels to readily measurable environmental parameters. J. Wat. Health 02.2:103-114.

Koplin, D.W., Furlong, E.T., Meyer, M.T., Thurman, E.M., Zaugg, S.D., Barber, L.B., and Buxton, H.T. 2002. Pharmaceuticals, hormones, and other organic wastewater contaminants in U.S. streams, 1999-2000: A national reconnaissance. Environ. Sci. Technol. 36:1202-1211.

Kuntz, R.L., Hartel, P.G., Rodgers, K., and Segars, W.I. 2004. Presence of *Enterococcus faecalis* in broiler litter and wild bird feces for bacterial source tracking. Water Research 38:3551-3557.

Leeming, R., Ball, A., Ashbolt, N., and Nichols, P. 1996. Using fecal sterols from humans and animals to distinguish fecal pollution in receiving waters. Water Res. 30:2893-2900.

Leeming, R., Nichols, P.D., and Ashbolt, N.J. 1998. Distinguishing sources of faecal pollution in Australian inland and coastal waters using sterol biomarkers and microbial faecal indicators. Water Services Association of Australia, No. 204, Melbourne.

Ley, V., Higgins, J., and Fayer, R. 2002. Bovine enteroviruses as indicators of fecal contamination. Appl. Environ. Microbiol. 68:3455-3461.

Long, S.C., El-Khoury, S.S., Oudejans, S.J.G., Sobsey, M.D., and Vinje, J. 2005. Assessment of sources and diversity of male-specific coliphages for source tracking. Environmental Engineering Science 22:367-377.

Maluquer de Motes, C., Clemente-Casares, P., Hundesa, A., Martin, M., and Girones, R. 2004. Detection of bovine and porcine adenoviruses for tracing the source of fecal contamination. Appl. Environ. Microbiol. 70:1448-1454.

Martellini, A., Payment, P., and Villemur, R. 2005. Use of eukaryotic mitochondrial DNA to differentiate human, bovine, porcine, and ovine sources in fecally contaminated surface water. Water Research 39:541-548.

Martin, B., Humbert, O., Camara, M., Guenzi, E., Walker, J., Mitchell, T., Andrew, P., Prudhomme, M., Alloing, G., Hakenbeck, R., Morrison, D.A., Boulnois, G.J., and Claverys, J.–P. 1992. A highly conserved repeated DNA element in the chromosome of *Streptococcus pneumoniae*. Nucl. Acids Res. 20:3479-3483.

Matsuki, T., Watanabe, K., Tanaka, R., and Oyaizu, H. 1998. Rapid identification of human intestinal bifidobacteria by 16S rRNA-targeted species and group-specific primers. FEMS Microbiol. Lett. 167:113-121.

McKeown, N.E., Fenaux, M., Halbur, P.G., and Meng, X.J. 2004. Molecular characterization of porcine TT virus, an orphan virus, in pigs from six different countries. Vet. Microbiol. 104:113-117.

McLauchlin, J., Amar, C., Pedraza-Diaz, S., and Nichols, G.L. 2000. Molecular epidemiological analysis of *Cryptosporidium* spp. in the United Kingdom: Results of genotyping *Cryptosporidium* spp. in 1,705 fecal samples from humans and 105 fecal samples from livestock animals. J. Clin. Microbiol. 38:3984-3990.

McLellan, S.L., Daniels, A.D., and Salmore, A.K. 2003. Genetic characterization of *Escherichia coli* populations from host sources of fecal pollution by using DNA fingerprinting. Appl. Environ. Microbiol. 69:2587-2594.

Moore, D.F., Harwood, V.J., Ferguson, D.M., Lukasik, J., Hannah, P., Getrich, M., and Brownwell, M. 2005. Evaluation of antibiotic resistance analysis and ribotyping for identification of faecal pollution sources in an urban watershed. J. Appl. Microbiol. 99:618-628.

Myoda, S.P., Carson, C.A., Fuhrmann, J.J., Hahm, B.–K., Hartel, P.G., Yampara-Iquise, H., Johnson, L., Kuntz, R.L., Nakatsu, C.H., Sadowsky, M.J., and Samadpour, M. 2003. Comparison of genotypic-based microbial source tracking methods requiring a host origin database. J. Wat. Health 1:167-180.

Nebra, Y., Bonjoch, X., and Blanch, A.R. 2003. Use of *Bifidobacterium dentium* as an indicator of the origin of fecal water pollution. Appl. Environ. Microbiol. 69:2651-2656.

Nichols, P.D., Leeming, R., Rayner, M.S., Latham, V., Ashbolt, N.J., and Turner, C. 1993. Comparison of the abundance of the fecal sterol coprostanol and fecal bacterial groups in inner-shelf waters and sediments near Sydney, Australia. J. Chromatog. 643:189-195.

Noble, R.T., Allen, S.M., Blackwood, A.D., Chu, W., Jiang, S.C., Lovelace, G.L., Sobsey, M.D., Stewart, J.R., and Wait, D.A. 2003. Use of viral pathogens and indicators to differentiate between human and non-human fecal contamination in a microbial source tracking comparison study. J. Wat. Health 1:195-207.

Noblet, J.A., Young, D.L., Zeng, E.Y., and Ensari, S. 2004. Use of fecal steroids to infer the sources of fecal indicator bacteria in the lower Santa Ana river watershed, California: Sewage is unlikely a significant source. Environ. Sci. Technol. 38:6002-6008.

Okamoto, H., Fukuda, M., Tawara, A., Nishizawa, T., Itoh, Y., Hayasaka, I., Tsuda, F., Tanaka, T., Miyakawa, Y., and Mayumi, M. 2000. Species-specific TT viruses and cross-species infection in nonhuman primates. J. Virol. 74:1132-1139.

Oshiro, R.K. and Olson, B.H. 1997. Occurrence of STh toxin gene in wastewater. In Coliforms and *E. coli*: Problem or Solution? Kay, D. and Fricker, C. (Eds.), The Royal Society of Chemistry, Cambridge.

Pacheco, A.B.F., Guth, B.E.C., Soares, K.C.C., Nishimura, L., Almeida, D.F., and Ferreira, L.C.S. 1997. Random amplification of polymorphic DNA reveals serotype-specific clonal clusters among enterotoxigenic *Escherichia coli* strains isolated from humans. J. Clin. Microbiol. 35:1521-1525.

Pang, X., Ding, D., Wei, G., Zhang, M., Wang, L., and Zhao, L. 2005. Molecular profiling of *Bacteroides* spp. in human feces by PCR-temperature gradient gel electrophoresis. J. Microbiol. Meth. 61:413-417.

Parveen, S., Portier, K.M., Robinson, K., Edmiston, L., and Tamplin, M. 1999. Discriminate analysis of ribotype profiles of *Escherichia coli* for differentiating human and nonhuman sources of fecal pollution. Appl. Environ. Microbiol. 65:3142-3147.

Parveen, S., Hodge, N.C., Stall, R.E., Farrah, S.R., and Tamplin, M.L. 2001. Genotypic and phenotypic characterization of human and non-human *Escherichia coli*. Water Res. 35:379-386.

Pond, K.R., Rangdale, R., Meijer, W.G., Brandao, J., Falcao, L., Rince, A., Masterson, B., Greaves, J., Gawler, A., McDonnell, E., Cronin, A.A., and Pedley, S. 2005. Workshop report: Developing pollution source tracking for recreational and shellfish waters. Environ. Forensics 5:237-247.

Power, M.L., Littlefield-Wyer, J., Gordon, D.M., Veal, D.A., and Slade, M.B. 2005. Phenotypic and genotypic characterization of encapsulated *Escherichia coli* isolated from blooms in two Australian lakes. Environ. Microbiol. 7:631-640.

Puglisi, E., Nicelli, M., Capri, E., Trevisan, M., and Del Re, A.A. 2003. Cholesterol, beta-sitosterol, ergosterol, and coprostanol in agricultural soils. J. Environ. Qual. 32:466-471.

Puig, A., Queralt, N., Jofre, J., and Araujo, R. 1999. Diversity of *Bacteroides fragilis* strains in their capacity to recover phages from human and animal wastes and from fecally polluted wastewater. Appl. Environ. Microbiol. 65:1772-1776.

Ram, J.L., Ritchie, R.P., Fang, J., Gonzales, F.S., and Selegean, J.P. 2004. Sequence-based source tracking of *Escherichia coli* based on genetic diversity of β-glucuronidase. J. Environ. Qual. 33:1024-1032.

Rangdale, R.E. 2003. Development of methods for pollution source tracking- Consolidated literature review. UK Environment Agency and the Centre for Environment, Fisheries and Aquaculture Sciences.

Ritter, K.J., Carruthers, E., Carson, C.A., Ellender, R.D., Harwood, V.J., Kingsley, K., Nakatsu, C., Sadowsky, M., Shear, B., West, B, Whitlock, J.E., Wiggins, B.A., and Wilbur, J.D. 2003. Assessment of statistical methods used in library-based approaches to microbial source tracking. J. Wat. Health 01.4:209-223.

Rochelle, P.A. 2002. *Giardia*: Detection and Occurrence in the Environment. In Encyclopedia of Environmental Microbiology, vol. 3. Bitton, G. (Ed.), pp. 1477-1489. John Wiley & Sons, Inc., New York.

Rousselon, N., Delgenes, J.-P., and Godon, J.-J. 2004. A new real time PCR (TaqMan PCR) system for detection of the 16S rDNA gene associated with fecal bacteria. J. Microbiol. Meth. 59:15-22.

Schaper, M., Duran, A.E., and Jofre, J. 2002a. Comparative resistance of phage isolates of four genotypes of F-specific RNA bacteriophages to various inactivation processes. Appl. Environ. Microbiol. 68:3702-3707.

Schaper, M., Jofre, J., Uys, M., and Grabow, W.O.K. 2002b. Distribution of genotypes of F-specific RNA bacteriophages in human and non-human sources of faecal pollution in South Africa and Spain. J. Appl. Microbiol. 92: 657-667.

Schvoerer, E., Ventura, M., Dubos, O., Cazaux, G., Serceau, R., Gournier, N., Dubois, V., Caminade, P., Fleury, H.J., and Lafon, M.E. 2001. Qualitative and quantitative molecular detection of enteroviruses in water from bathing areas and from a sewage treatment plant. Res. Microbiol. 152:179-186.

Scott, T.M., Rose, J.B., Jenkins, T.M., Farrah, S.R., Lukasik, J. 2002. Microbial source tracking: Current methodology and future directions. Appl. Environ. Microbiol. 68:5796-5803.

Scott, T.M., Parveen, S., Portier, K.M., Rose, J.B., Tamplin, M.L., Farrah, S.R., Koo, A., and Lukasik, J. 2003. Geographical variation in ribotype profiles of *Escherichia coli* isolates from humans, swine, poultry, beef and dairy cattle in Florida. Appl. Environ. Microbiol. 69:1089-1092.

Scott, T.M., Jenkins, T.M., Lukasik, J., and Rose, J.B. 2005. Potential use of a host-associated molecular marker in *Enterococcus faecium* as an index of human fecal pollution. Environ. Sci. Technol. 39:283-287.

Seurinck, S., Verstraete, W., and Siciliano, S.D. 2003. Use of 16S-23S rRNA intergenic spacer region PCR and repetitive extragenic palindromic PCR analyses of *Escherichia coli* isolates to identify nonpoint fecal sources. Appl. Environ. Microbiol. 69:4942-4950.

Simpson, J.M., Santo Domingo, J.W., and Reasoner, D.J. 2002. Microbial source tracking: State of the science. Environ. Sci. Technol. 36:5279-5288.

Sinton, L.W., Finlay, R.K., and Hannah, D.J. 1998. Distinguishing human from animal fecal contamination in water: A review. N. Z. J. Mar. Freshwater Res. 32:323-348.

Sirivedhin, T. and Gray, K.A. 2005. Identifying anthropogenic markers in surface waters influenced by treated effluents: a tool in potable water reuse. Wat. Res. 39:1154-1164.

Souza, V., Rocha, M., Valera, A., and Eguiarte, L.E. 1999. Genetic structure of natural populations of *Escherichia coli* in wild hosts on different continents. Appl. Environ. Microbiol. 65:3373-3385.

Stern, M.J., Ames, G.F.–L., Smith, N.H., Robinson, E.C., and Higgins, C.F. 1984. Repetitive extragenic palindromic sequences: a major component of the bacterial genome. Cell 37:1015-1026.

Stewart, J.R., Ellender, R.D., Gooch, J.A., Jiang, S., Myoda, S.P., and Weisberg, S.B. 2003. Recommendations for microbial source tracking: Lessons learned from a methods comparison study. J. Water Health 1.4:225-231.

Stoeckel, D.M., Mathes, M.V., Hyer, K.E., Hagedorn, C., Kator, H., Lukasik, J., O'Brien, T.L., Fenger, T.W., Samadpour, M., Strickler, K.M., and Wiggins, B.A. 2004. Comparison of seven protocols to identify fecal contamination sources using *Escherichia coli*. Environ. Sci. Technol. 38:6109-6117.

Sulaiman, I.M., Xiao, L., Yang, C., Escalante, L., Moore, A., Beard, C.B., Arrowood, M.J., and Lal, A.A. 1998. Differentiating human from animal isolates of *Cryptosporidium parvum*. Emerg. Infect. Dis. 4:681-685.

Suprihatin, I., Fallowfield, H., Bentham, R., and Cromar, N. 2003. Determination of faecal pollutants in Torrens and Patawalonga catchment waters in South Australia using faecal sterols. Wat. Sci. Technol. 47:283-289.

Swaminathan, B., Barrett, T.J., Hunter, S.B., and Tauxe, R.V. 2001. PulseNet: The molecular subtyping network for foodborne bacterial disease surveillance, United States. Emerg. Infect. Dis. 7:382-389.

U.S. Environmental Protection Agency. 2005. Microbial Source Tracking Guide Document. EPA/600-R-05-064, Office of Research and Development, Cincinnati, OH.

Vankerckhoven, V., Van Autgaerden, T., Vael, C., Lammens, C., Chapelle, S., Rossi, R., Jabes, D., and Goossens, H. 2004. Development of a multiplex PCR for the detection of *asa1*, *gelE*, *cylA*, *esp*, and *hyl* genes in enterococci and survey for virulence determinants among European hospital isolates of Enterococcus faecium. J. Clin. Microbiol. 42:4473-4479.

Venieri, D., Vantarakis, A., Komninou, G., and Papapetropoulou, M. 2004. Differentiation of faecal Escherichia coli from human and animal sources by random amplified polymorphic DNA-PCR (RAPD-PCR). Wat. Sci. Technol. 50:193-198.

Versalovic, J., Koeuth, T., and Lupski, J.R. 1991. Distribution of repetitive DNA sequences in eubacteria and application to fingerprinting of bacterial genomes. Nucl. Acids. Res. 19:6823-6831.

Vinje, J., Oudejans, S.J.G., Stewart, J.R., Sobsey, M.D., and Long, S.C. 2004. Molecular detection and genotyping of male-specific coliphages by reverse transcription-PCR and reverse line blot hybridization. Appl. Environ. Microbiol. 70:5996-6004.

Wallis, J.L. and Taylor, H.D. 2003. Phenotypic population characteristics of the enterococci in wastewater and animal faeces: Implications for the new European directive on the quality of bathing waters. Water Sci. Technol. 47:27-32.

Wheeler, A.L., Hartel, P.G., Godfrey, D.G., Hill, J.L., and Segars, W.I. 2002. Potential of *Enterococcus faecalis* as a human fecal indicator for microbial source tracking. J. Environ. Qual. 31:1286-1293.

Whiting-Grant, K., Dalton, C., and Dillon, F. 2003. Microbial source tracking in two Southern Maine watersheds. Webhannet River Watershed Report Number MSG-TR-03-1, Maine Sea Grant Project.

Wiggins, B.A., Andrews, R.W., Conway, R.A., Corr, C.L., Dobratz, E.J., Dougherty, D.P., Eppard, J.R., Knupp, S.R., Limjoco, M.C., Mettenburg, J.M., Rinehardt, J.M., Sonsino, J., Torrijos, R.L., and Zimmerman, M.E. 1999. Use of antibiotic resistance analysis to identify nonpoint sources of fecal pollution. Appl. Environ. Microbiol. 65:3483-3486.

Wiggins, B.A., Cash, P.W., Creamer, W.S., Dart, S.E., Garcia, P.P., Gerecke, T.M., Han, J., Henry, B.L., Hoover, K.B., Johnson, E.L., Jones, K.C., McCarthy, J.G., McDonough, J.A., Mercer, S.A., Noto, M.J., Park, H., Philips, M.S., Purner, S.M., Smith, B.M., Stevens, E.N., and Varner, A.K. 2003. Use of antibiotic resistance analysis for representativeness testing of multiwatershed libraries. Appl. Environ. Microbiol. 69:3399-3405.

Yang, H.–S., Vinopal, R.T., Grasso, D., and Smets, B.F. 2004. High diversity among environmental *Escherichia coli* isolates from a bovine feedlot. Appl. Environ. Microbiol. 70:1528-1536.

Xiao, L., Alderisio, K., Limor, J., Royer, M., and Lal, A.A.. 2000. Identification of species and sources of *Cryptosporidium* oocysts in storm waters with a small-subunit rRNA-based diagnostic and genotyping tool. Appl. Environ. Microbiol. 66:5492-5498.

Xiao, L., Bern, C., Limor, J., Sulaiman, I., Roberts, J., Checkley, W., Cabrera, I., Gilman, R.H., and Lal, A.A. 2001. Identification of 5 types of *Cryptosporidium* parasites in children from Lima, Peru. J. Infect. Dis. 185:270-271.

Zuccato, E., Chiabrando, C., Castiglioni, S., Calamari, D., Bagnati, R., Schiarea, S., and Fanelli, R. 2005. Cocaine in surface waters: A new evidence-based tool to monitor community drug abuse. Environ. Health: A Global Access Science Source 4:14.

WASTEWATER UTILITY

Alabama
Montgomery Water Works & Sanitary Sewer Board

Alaska
Anchorage Water & Wastewater Utility

Arizona
Gila Resources
Glendale, City of, Utilities Department
Mesa, City of
Peoria, City of
Phoenix Water Services Department
Pima County Wastewater Management

Arkansas
Little Rock Wastewater Utility

California
Calaveras County Water District
Central Contra Costa Sanitary District
Corona, City of
Crestline Sanitation District
Delta Diablo Sanitation District
Dublin San Ramon Services District
East Bay Dischargers Authority
East Bay Municipal Utility District
Eastern Municipal Water District
El Dorado Irrigation District
Fairfield-Suisun Sewer District
Fresno Department of Public Utilities
Inland Empire Utilities Agency
Irvine Ranch Water District
Las Virgenes Municipal Water District
Livermore, City of
Lodi, City of
Los Angeles, City of
Napa Sanitation District
Orange County Sanitation District
Palo Alto, City of
Riverside, City of
Sacramento Regional County Sanitation District
San Diego Metropolitan Wastewater Department, City of
San Francisco, City & County of
Sanitation Districts of Los Angeles County
San Jose, City of
Santa Barbara, City of
Santa Cruz, City of
Santa Rosa, City of
South Bayside System Authority
South Coast Water District
South Orange County Wastewater Authority
Stege Sanitary District
Sunnyvale, City of
Union Sanitary District
West Valley Sanitation District

Colorado
Aurora, City of
Boulder, City of
Colorado Springs Utilities
Greeley, City of
Littleton/Englewood Water Pollution Control Plant
Metro Wastewater Reclamation District, Denver

Connecticut
The Mattabassett District
Greater New Haven WPCA

District of Columbia
District of Columbia Water & Sewer Authority

Florida
Broward, County of
Fort Lauderdale, City of
JEA
Miami-Dade Water & Sewer Authority
Orange County Utilities Department
Reedy Creek Improvement District
Seminole County Environmental Services
St. Petersburg, City of
Stuart Public Utilities
Tallahassee, City of
Tampa, City of
Toho Water Authority
West Palm Beach, City of

Georgia
Atlanta Department of Watershed Management
Augusta, City of
Clayton County Water Authority
Cobb County Water System
Columbus Water Works
Fulton County
Gwinnett County Department of Public Utilities
Savannah, City of

Hawaii
Honolulu, City & County of

Idaho
Boise, City of

Illinois
American Bottoms Wastewater Treatment Plant
Greater Peoria Sanitary District
Kankakee River Metropolitan Agency
Metropolitan Water Reclamation District of Greater Chicago
Wheaton Sanitary District

Iowa
Ames, City of
Cedar Rapids Wastewater Facility
Des Moines, City of
Iowa City

Kansas
Johnson County Unified Wastewater Districts
Unified Government of Wyandotte County/ Kansas City, City of

Kentucky
Louisville & Jefferson County Metropolitan Sewer District

Louisiana
Sewerage & Water Board of New Orleans

Maine
Bangor, City of
Portland Water District

Maryland
Anne Arundel County Bureau of Utility Operations
Hagerstown, City of
Howard County Department of Public Works
Washington Suburban Sanitary Commission

Massachusetts
Boston Water & Sewer Commission
Upper Blackstone Water Pollution Abatement District

Michigan
Ann Arbor, City of
Detroit, City of
Holland Board of Public Works
Lansing, City of
Saginaw, City of
Wayne County Department of Environment
Wyoming, City of

Minnesota
Rochester, City of
Western Lake Superior Sanitary District

Missouri
Independence, City of
Kansas City Missouri Water Services Department
Little Blue Valley Sewer District
Metropolitan St. Louis Sewer District

Nebraska
Lincoln Wastewater System

Nevada
Henderson, City of
Reno, City of

New Jersey
Bergen County Utilities Authority
Ocean, County of
Passaic Valley Sewerage Commissioners

New York
New York City Department of Environmental Protection

North Carolina
Charlotte/Mecklenburg Utilities
Durham, City of
Metropolitan Sewerage District of Buncombe County
Orange Water & Sewer Authority

Ohio
Akron, City of
Butler County Department of Environmental Services
Columbus, City of
Metropolitan Sewer District of Greater Cincinnati
Northeast Ohio Regional Sewer District
Summit, County of

Oklahoma
Tulsa, City of

Oregon
Clean Water Services
Eugene/Springfield Water Pollution Control
Water Environment Services

Pennsylvania
Philadelphia, City of
University Area Joint Authority

South Carolina
Charleston Commissioners of Public Works
Mount Pleasant Waterworks & Sewer Commission
Spartanburg Sanitary Sewer District

Tennessee
Cleveland, City of
Knoxville Utilities Board
Murfreesboro Water & Sewer Department
Nashville Metro Water Services

Texas
Austin, City of
Dallas Water Utilities
Denton, City of
El Paso Water Utilities
Fort Worth, City of
Gulf Coast Waste Disposal Authority
Houston, City of
San Antonio Water System
Trinity River Authority

Utah
Salt Lake City Corporation

Virginia
Alexandria Sanitation Authority

Arlington, County of
Fairfax County
Hampton Roads Sanitation District
Henrico, County of
Hopewell Regional Wastewater Treatment Facility
Loudoun County Sanitation Authority
Lynchburg Regional WWTP
Prince William County Service Authority
Richmond, City of
Rivanna Water & Sewer Authority

Washington
Everett, City of
King County Department of Natural Resources
Seattle Public Utilities
Sunnyside, Port of
Yakima, City of

Wisconsin
Green Bay Metro Sewerage District
Kenosha Water Utility
Madison Metropolitan Sewerage District
Milwaukee Metropolitan Sewerage District
Racine, City of
Sheboygan Regional Wastewater Treatment
Wausau Water Works

Australia
South Australian Water Corporation
Sydney Water Corporation
Water Corporation of Western Australia

Canada
Greater Vancouver Regional District
Regina, City of, Saskatchewan
Toronto, City of, Ontario
Winnipeg, City of, Manitoba

New Zealand
Watercare Services Limited

United Kingdom
Yorkshire Water Services Limited

STORMWATER UTILITY

California
Los Angeles, City of, Department of Public Works
Monterey, City of
San Francisco, City & County of
Santa Rosa, City of

Colorado
Aurora, City of
Boulder, City of

Georgia
Griffin, City of

Iowa
Cedar Rapids Wastewater Facility
Des Moines, City of

Kansas
Overland Park, City of

Maine
Portland Water District

Minnesota
Western Lake Superior Sanitary District

North Carolina
Charlotte, City of, Stormwater Services

Pennsylvania
Philadelphia, City of

Tennessee
Chattanooga Stormwater Management

Washington
Bellevue Utilities Department
Seattle Public Utilities

STATE

Arkansas Department of Environmental Quality
Fresno Metropolitan Flood Control District
Kansas Department of Health & Environment
Kentucky Department of Environmental Protection
Ohio River Valley Sanitation Commission
Urban Drainage & Flood Control District, CO

CORPORATE

Abt. Associates
ADS Environmental Services
Alan Plummer & Associates
Alden Research Laboratory Inc.
American Water
Aqua-Aerobic Systems Inc.
Aquateam–Norwegian Water Technology Centre A/S
ARCADIS
Associated Engineering
Black & Veatch
Blasland, Bouck & Lee Inc.
Boyle Engineering Corporation
Brown & Caldwell
Burns & McDonnell
CABE Associates Inc.
The Cadmus Group
Camp Dresser & McKee Inc.
Carollo Engineers Inc.
Carpenter Environmental Associates Inc.
CDS Technologies Inc.
Chemtrac Systems Inc.
CH2M HILL
Construction & Environmental Consulting, Inc.
D&B/Guarino Engineers, LLC
Damon S. Williams Associates, LLC
Dewling Associates Inc.
Earth Tech Inc.
Ecovation
EMA Inc.
Environ/The ADVENT Group, Inc.
The Eshelman Company Inc.
EWT Holdings Corporation
Fay, Spofford, & Thorndike Inc.
Freese & Nichols Inc.
ftn Associates Inc.
Fuss & O'Neill Inc.
Gannett Fleming Inc.
Geosyntec Consultants
GHD
Golder Associates Ltd.
Greeley and Hansen LLC
Hazen & Sawyer, P.C.
HDR Engineering Inc.
HNTB Corporation
HydroQual Inc.
Infilco Degremont Inc.
Jacobson Helgoth Consultants Inc.
Jason Consultants LLC Inc.
Jordan, Jones, & Goulding Inc.
KCI Technologies Inc.
Kelly & Weaver, P.C.
Kennedy/Jenks Consultants
KMK Consultants
Komline Sanderson Engineering Corporation
Lawler, Matusky & Skelly Engineers, LLP
Limno-Tech Inc.
Lombardo Associates Inc.
Malcolm Pirnie Inc.
Material Matters
McKim & Creed
MEC Analytical Systems Inc.
Metcalf & Eddy Inc.
MPR Engineering Corporation, Inc.
MWH
New England Organics
O'Brien & Gere Engineers Inc.
Odor & Corrosion Technology Consultants Inc.
Oswald Green, LLC
PA Government Services Inc.
Parametrix Inc.
Parsons
Post, Buckley, Schuh & Jernigan
R&D Engineering/Conestoga Rover & Associates
RMC Inc.
R.M. Towill Corporation
Ross & Associates Ltd.
Rothberg, Tamburini & Windsor, Inc.
Royce Technologies
Stantec Consulting Inc.
Stearns & Wheler, LLC
Stone Environmental Inc.
Stormwater360
Stratus Consulting Inc.
Synagro Technologies Inc.
Tetra Tech Inc.
Trojan Technologies Inc.
Trussell Technologies, Inc.
URS Corporation
USfilter
Wade-Trim Inc.
Westin Engineering Inc.
Weston Solutions Inc.
Woodard & Curran
WRc/D&B, LLC
WWETCO, LLC
Zenon Environmental Inc.
Zoeller Pump Company

INDUSTRY

American Electric Power
ChevronTexaco Energy Research & Technology Company
The Coca-Cola Company
Dow Chemical Company
DuPont Company
Eastman Chemical Company
Eastman Kodak Company
Eli Lilly & Company
Merck & Company Inc.
Premier Chemicals LLC
Procter & Gamble Company
RWE Thames Water Plc
Severn Trent Services Inc.
Suez Environnment
United Water Services LLC

Note: List as of 4/1/06

IWA Publishing's authorised EU representative for General Product Safety Regulations is Diane D'Arras, 15 rue Duret, 75116 Paris, France, e-mail: safety@iwap.co.uk.

Printed and bound by CPI Group (UK) Ltd, Croydon, CR0 4YY
12/05/2026
02108892-0010